CLOUD PHYSICS

CLOUD PHYSICS

By

D. W. PERRIE

Meteorologist, British Columbia Forest Service

UNIVERSITY OF TORONTO PRESS: 1950

JOHN WILEY & SONS, INC.

NEW YORK: 1950

Foreword

IN THESE times of rapidly increasing interest in cloud physics the appearance of this very readable book on the subject fills an immediate need. Everybody looks at the clouds, usually to judge whether the sun will come out or whether the clouds will thicken and yield rain or snow, but often simply to enjoy their beauty, especially when interesting contours or brilliant colourings are involved. Meteorologists view clouds as a professional tool, an indication of what is going on aloft, free for the looking. Aviators, who must fly among or through the clouds, have a very personal interest in their location and characteristics; they ask, how high, how thick, how rough, how icy? And now water-supply engineers have begun to investigate the practicability of making clouds yield rain to order. Whether one looks at the clouds for pleasure, for forecasting, for flying, or for making rain, a knowledge of cloud physics enhances the enjoyment or improves the applications.

Mr. Perrie is particularly well qualified to provide such knowledge,— weighing and summarizing what others have done and adding his own. He is a well-trained physicist-meteorologist especially interested in clouds. He has observed them from the ground; he has photographed them; he has used them in forecasting and, in turn, has forecast their transformations, and he has flown up and seeded clouds with dry ice to induce precipitation. His competence as a cloud specialist is attested by the fact that he is one of the members of the Committee for the Study of Clouds and Hydrometeors of the International Meteorological Organization.

Although the title is "Cloud Physics," the book is more than this. Clouds are described and named, and the large-scale processes responsible for the various formations are explained. Then the internal physics comes in for discussion: nuclei, how rain is formed naturally, and how additional rain can be induced artificially. A substantial portion of the book is reserved, however, for other cloud topics of interest: observing, forecasting, flying, and, at the end, the intriguingly rare nacreous and noctilucent clouds, high in the stratosphere, and optical and electrical phenomena. A list of references for the specialist and a convenient glossary for the layman complete a well-rounded and satisfying book.

CHARLES F. BROOKS
Professor of Meteorology, and Director

Blue Hill Meteorological Observatory,
Harvard University,
Milton, 86, Mass., U.S.A.,
April 23, 1950.

Acknowledgments

For many helpful suggestions and much encouragement, I wish to thank sincerely Mr. J. R. H. Noble, who read the manuscript with great care. I am also deeply indebted to Mr. Andrew Thomson, Mr. F. W. Benum, and Dr. C. F Brooks for valuable suggestions. The encouragement of Dr. John Patterson and Mr. A. D. Thiessen is also gratefully acknowledged.

Photographs have been very kindly provided by individuals and organizations, as acknowledged in the captions. In this connection, I am much indebted to Mr. I. R. Tannehill for his aid in obtaining photographs from the collection of the United States Weather Bureau.

D. W. P.

Contents

PLATES

I. Cirrus in more or less parallel trails and small patches.
United States Weather Bureau—F. Ellerman.

II. Cirrocumulus and cirrus. *Author.*

III. Altocumulus. The patch on the right has a very interesting wave form. *United States Weather Bureau.*

IV. Artificial cloud produced by the passage of an aircraft. *United States Weather Bureau—U.S. Army.*

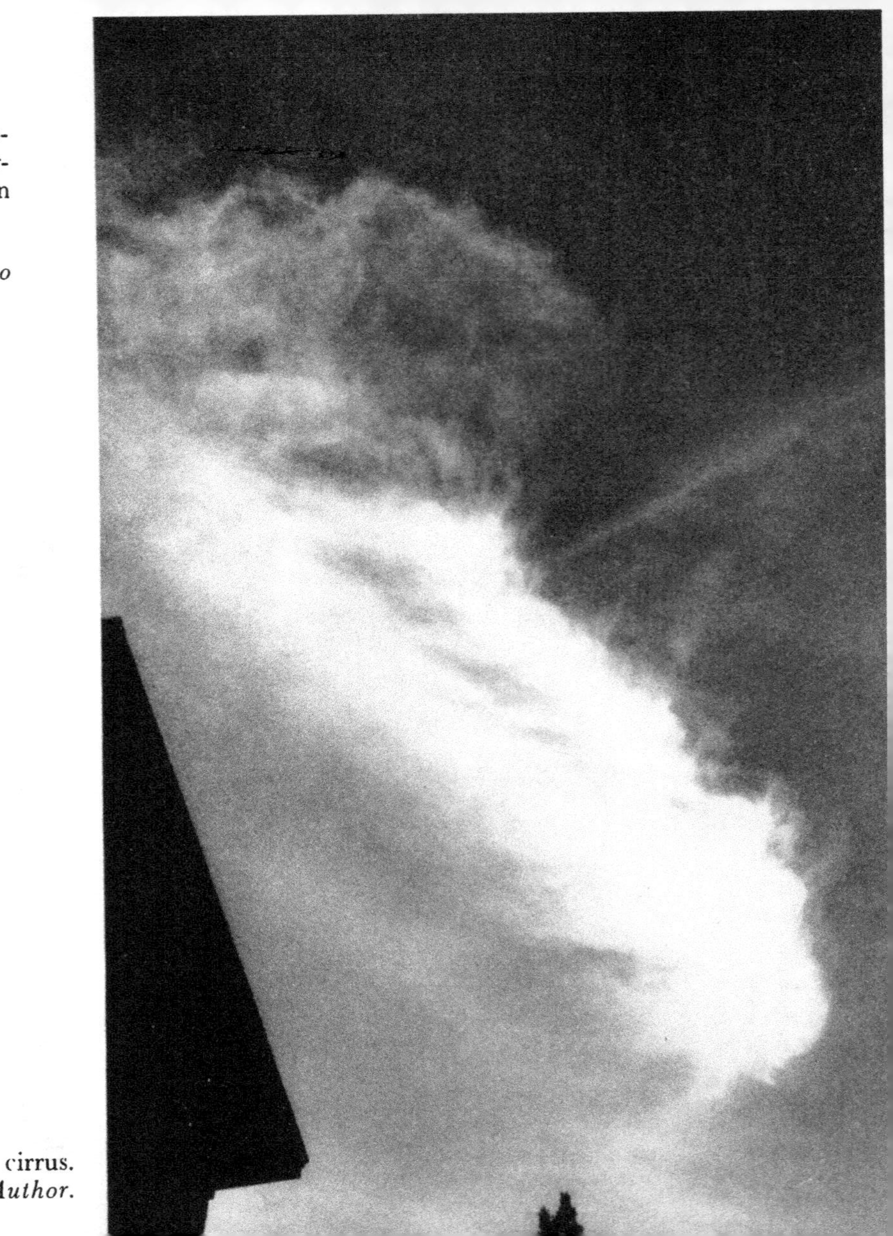

V. Cirrus, with part of the campus of the University of Alberta in the foreground.
Courtesy the New Trail—*photo by McDermid Studios.*

VI. Dense cirrus.
Author.

VII. Altocumu-lus or high strato-cumulus, at one level. *United States Weather Bureau—A. C. Lapsley.*

VIII. Telephoto view of Half Dome mountain enveloped in clouds. *United States National Park Service.*

IX. Altocumulus and altostratus, with fractostratus and stratocumulus below. *Mrs. James A. Selby.*

X. Stratocumulus clouds over the Saskatchewan River. There is a significant break over the river itself, caused by a downdraft over the water. *Royal Canadian Air Force.*

XI. Top of a layer of stratocumulus, with mountain tops projecting above the cloud. *Trans-Canada Air Lines.*

XII. Patches of stratocumulus and fractostratus, as seen from above. *Trans-Canada Air Lines.*

XIII. Heavy cumulus, as seen from the ground.
Mrs. James A. Selby.

XIV. Heavy cumulus, as seen from an aircraft near the
cloud tops. *American Airlines.*

XV. Cumulonimbus. An anvil is partly formed, and the
hard outlines have dissolved in the upper portions
of the cloud mass. There is a small pileus to the left
of the summit. A shower is visible below the cloud.
United States Weather Bureau—Harold T. Floreen.

XVI. Cirrus of a cumulonimbus anvil, frayed out at upper
left. *United States Weather Bureau—Harold T. Floreen.*

XVII. A magnificent example of a fully developed cumulonimbus, with an anvil head. *A. F. McQuarrie.*

XVIII. A typical top of a fine weather cumulus. *Author.*

XIX. Fine weather cumulus, and stratocumulus formed
from cumulus. *E. D. M. Williams.*

XX. Fine weather cumulus, and some fractocumulus.
T.W.A. Airlines.

XXI. Lenticular altocumulus. *United States Weather Bureau—Maxwell Parshall.*

XXII. Crepuscular rays, with cumulus and stratocumulus. *United States Weather Bureau.*

XXIII. "Table cloth" cloud, on Table Mountain, Cape Town, South Africa. *United States Weather Bureau.*

XXIV. Crest cloud on the Rock of Gibralter. This cloud stretches out a mile or more when an east wind, known as the "levanter" is blowing. The cloud is known locally as the Levanter cloud. *British Meteorological Office.*

XXV. Tenuous fractostratus formed from lifting fog in the Gatineau River valley. This condition persisted for a short time in the early morning. *Author*.

XXVI. The mammilated base of thunder cloud. *United States Weather Bureau—Harold Photograph Studio*.

XXVII. Nacreous or mother-of-pearl clouds, photographed after sunset, at Oslo, Norway, by Professor Carl Störmer. *United States Weather Bureau—Carl Störmer.*

XXVIII. Noctilucent cloud. *United States Weather Bureau—Carl Störmer.*

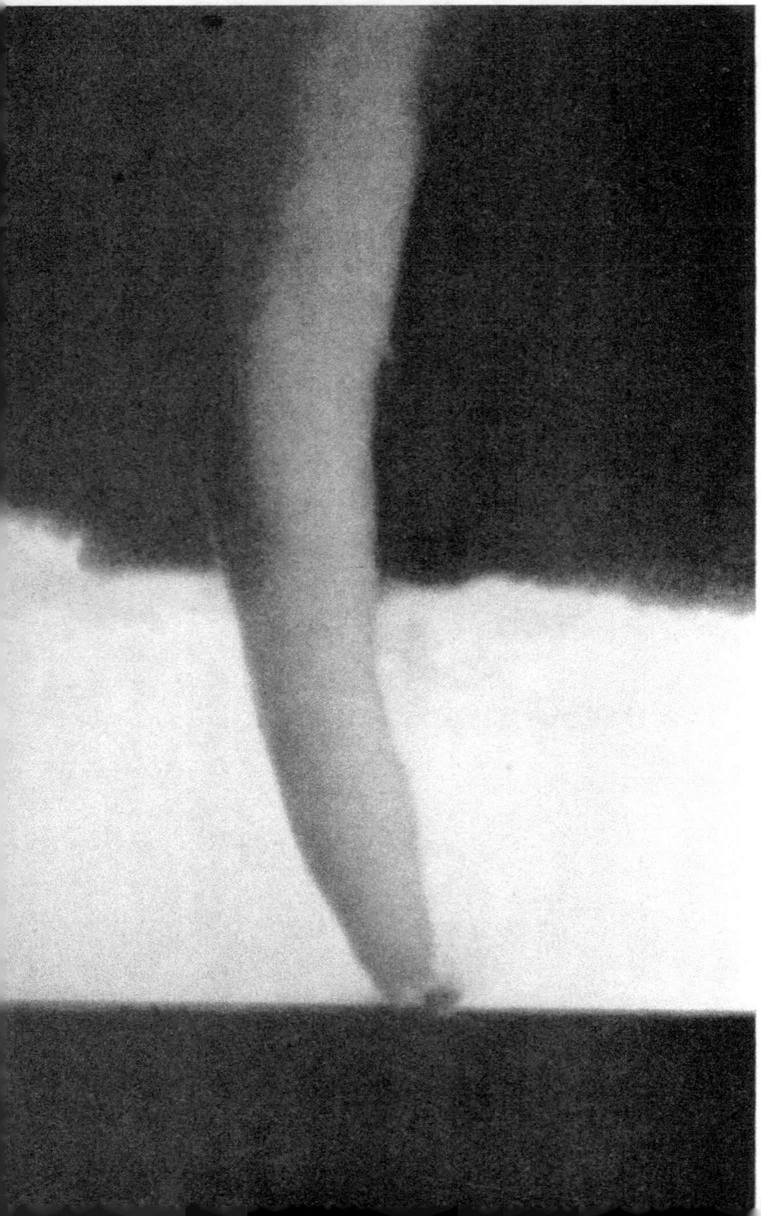

XXIX. Tornado, showing the typical funnel cloud. *United States Weather Bureau* –I. B. Blackstock.

XXX. Lightning. *United States Weather Bureau* –Charles S. Watson.

CLOUD PHYSICS

1. Forms and Classification

To THE casual observer, a cloud appears to be a mass or group of masses of some visible substance, floating in the air above the earth. A cloud is, however, made up of an aggregation of a great number of very small "elements." These "elements" are water droplets or ice crystals, or a mixture of water droplets and ice crystals. Clouds are, in general, supported by vertical currents in the atmosphere, and, although they appear to float, their elements are falling relative to the surrounding air.

Clouds assume an almost unlimited number of different forms, and for convenience it is necessary that some system of classification be used. The first step in this direction was taken by the French naturalist Lamarck, who published a cloud classification in 1802. He pointed out that his object was to describe the principal forms of cloud and not to describe the material of the clouds. Lamarck's classification comprised five principal cloud types, with some subdivisions. The terminology employed was French and the work was not illustrated. This classification received very little notice.

In 1803, Luke Howard published a classification under the title, *On the Modifications of Clouds.* The terminology was Latin, and the work was illustrated. Howard's fundamental cloud types became the basis for future cloud classification. There was an interesting similarity between Howard's classification and Lamarck's, four of Lamarck's five principal types being recognizable. In 1840, another type was recognized when Kaemtz introduced the term "stratocumulus."

In the later portion of the nineteenth century, several notable contributions were made. An attempt was made by Renou in 1855 to group the clouds into two fundamental categories: divided forms, and spread-out or veil forms. He also introduced "middle" clouds, between the high clouds and the low clouds. New criteria, notably considerations of altitude, were introduced into Howard's classification. These altitude distinctions were favoured by Hildebrandsson, who thought of applying them to the determination of winds at different levels. A distinction between ice clouds and liquid water clouds was made by Poey. The form "cumulonimbus" was introduced by Weilbach.

Notable work on clouds was contributed by the Reverend Clement Ley and the Honourable Ralph Abercromby. The latter was much interested in the similarity of cloud types throughout the world, and made two journeys around the globe to satisfy himself that the similarity existed. Ley was particularly concerned with the castellated forms, to which he applied the name "turreted stratus," and of which he spoke as "an especial favourite of my own."

At the International Conference at Munich in 1891 a classification was agreed upon, and as a result the first *International Cloud Atlas* was published in 1896. The meteorologists most active in this work were Hildebrandsson, Teisserenc de Bort, and Riggenbach. Several editions of the *International Cloud Atlas* appeared, the last being published in 1910.

In 1922, an International Commission for the Study of Clouds was set up to prepare a new Cloud Atlas. The Commission was headed by General E. Delcambre, Director of the Office Nationale Météorologique de France. In 1930 the Commission published an abridged edition of the Atlas, and in 1932 the complete Atlas, which was produced in English, French, German, and Spanish editions.

In 1933, after the publication of the Atlas, it was decided at a meeting of the International Meteorological Committee at De Bilt that the term "nimbus" should be abandoned and that the term "fractonimbus" might be adopted for low ragged clouds of bad weather, although the use of the terms "fractostratus" and "fractocumulus" was recommended.

The observed cloud forms adapt themselves to separation into fairly definite groups. Clouds occur generally in "heap" forms or "sheet" or "layer" forms, or combinations of these. The term "veil clouds" is very aptly used at times for certain of the sheet clouds, particularly the precipitating forms. Clouds are grouped in four families, high clouds, middle clouds, low clouds, and clouds of vertical development. All cloud heights mentioned in the literature are referred to the general level of the terrain in the vicinity.

HIGH CLOUDS

High clouds usually occur at heights from 20,000 to 40,000 feet above the terrain, but may be found in high latitudes at much lower levels.

Cirrus clouds are separate clouds, delicate and fibrous. They are generally white, often silky, and do not usually have any shading. The appearance of cirrus varies considerably, taking the forms of tufts, lines, or plumes. The effect of perspective causes lines of cirrus to appear to converge to a point on the horizon.

Cirrus occurs mostly in forms free from shadows, but there is a notable exception in the dense cirrus which results from the breaking away of the tops of cumulonimbus clouds. Owing to their slight density and their thinness, cirrus clouds do not dim the outlines of the sun or moon, and sometimes do not even make a really appreciable change in the appearance of the sky.

The colours of cirrus deviate occasionally from pure white. When the sun is low, colours of yellow or red sometimes show and at other times a dark grey colour is observed.

Isolated wisps of snow, seen against a blue sky, sometimes bear a considerable resemblance to cirrus. On closer examination they are revealed to be of a less pure white and to lack the silky texture of cirrus. Thin wisps of rain should be easily distinguishable from cirrus, on account of their grey colour. These wisps of rain will also sometimes display rainbows or portions of them, which cirrus clouds cannot possibly do.

Cirrus filosus consists of filaments which may be curved, or almost straight. However, neither tufts nor small points belong in this group, nor may any of the parts be fused together.

Cirrus uncinus is in the shape of commas, the upper parts being pointed or ending in little tufts.

Cirrus densus clouds are so thick that they may be mistaken for middle or low clouds. This cirrus may exhibit shadows. It is often formed as a result of the anvils breaking away from cumulonimbus clouds.

Cirrus nothus comes from cumulonimbus and is essentially the same as cirrus densus. It consists of recently broken-away cumulonimbus anvils. In such clouds the anvil form many sometimes still be seen. The most satisfactory name for cirrus formed from cumulonimbus heads is *tonitro-cirrus,* or "thunder cirrus." The name "false cirrus" is sometimes used for this and for the ice crystal head while still attached to the cumulonimbus. This term is unfortunate, since the ice crystal head is part of the cumulonimbus, and when detached becomes true cirrus.

Cirrostratus is a whitish veil of cloud, so thin that it does not blur the outlines of the sun or moon. It commonly causes haloes. It has, at times, a fibrous structure. Cirrostratus may be very diffuse, and merely give the sky a milky look. Parhelia are quite commonly observed in cirrostratus, and paraselenae are also occasionally observed.

Cirrocumulus consists of small flakes or very small globular elements, arranged in various ways, as groups or lines, or as ripples resembling sand on the seashore. It does not usually have shadows. It is generally considered that cirrocumulus is a degraded state of cirrus or cirrostratus, both of which may change into it; and that in such cases the cirrocumulus retains some of the fibrous characteristics of the cirrus or cirrostratus. This does not appear to be invariably true. Cirrocumulus sometimes is observed when a watcher on the ground has been unable to see cirrus or cirrostratus. There is always the possibility that cirrus or cirrostratus may have been present, but in a form too thin to be observed from the ground. In any case, real cirrocumulus is uncommon.

Coronae and irisations are occasionally observed in cirrocumulus, although they are generally considered characteristic of altocumulus.

MIDDLE CLOUDS

Middle clouds usually occur, in middle latitudes, at heights of 6,000 to 20,000 feet. They are found at heights near the lower limit of this range in the colder seasons, and at heights near the upper limit in the warmer seasons.

Altocumulus is made up of laminae, or of flattened globules, the smallest of which may be very small and thin. Altocumulus is frequently shaded. The elements are commonly arranged in groups, lines, or waves; sometimes with their edges joined. Altocumulus occurs in a great variety of forms, and not uncommonly in several levels at the same time. Irisations are quite often observed in the thin edges of altocumulus, and are considered characteristic of this cloud form. They also occur in cirrocumulus, and so are not exclusively altocumulus phenomena. Coronae are fairly common in altocumulus. The presence of a corona is sometimes used as an argument that the cloud observed is altocumulus, and not stratocumulus or cirrocumulus. This argument is invalid, however, for coronae are by no means rare in stratocumulus and also occur in cirrocumulus.

There are a considerable number of borderline cases where it is difficult to distinguish positively between altocumulus and the higher forms of stratocumulus on the one hand, and between altocumulus and the lower forms of cirrocumulus on the other.

Appreciable amounts of precipitation do not usually fall from altocumulus. In most cases the precipitation is confined to virga, that is, trailing precipitation which evaporates and so does not reach the ground. In combination with altostratus, however, altocumulus sometimes forms part of a precipitating cloud. This altostratus-altocumulus combination is fairly common when there is overrunning warm air.

Altocumulus translucidus has thin spots and interstices between elements, the interstices and thin spots letting through considerable sunlight or moonlight.

Altocumulus opacus is thick and rather dense. This form occurs most frequently in combination with altostratus.

Altocumulus cumulogenitus is formed from cumulus clouds by the spreading of the tops, and dissolution of the bases. This change usually occurs towards evening.

Altocumulus floccus takes the form of considerable numbers of small elements, with appreciable vertical development. It is of frequent occurrence.

Altostratus is a veil of grey or bluish cloud, which appears striated or fibrous. The thinner forms of altostratus differ little from the thicker forms of cirrostratus. The sun or moon shows vaguely through altostratus, much as though it were viewed through ground glass. The thicker forms hide the sun or moon completely, although relatively light patches may be observed between dark parts. The striated or fibrous structure shows in places, notably near the sun or moon. Relief is not observed on the base of altostratus, although irregularities of the base occur, often caused by the presence of virga.

Altostratus translucidus is the form of altostratus which first appears when cirrostratus thickens. It is a light grey, and the sun or moon may be seen as through ground glass.

Altostratus opacus is an opaque layer of variable thickness which shows a fibrous structure in parts. It may, at least in the thicker spots, completely hide the sun or moon.

Altostratus precipitans is a layer of opaque altostratus from which rain or snow falls. The rain will generally be light, and the snow moderate. The first precipitation from altostratus precipitans is in the form of virga. The wet appearance imparted to the cloud base by the virga is sufficient to distinguish the cloud as altostratus precipitans.

LOW CLOUDS

Low clouds occur at heights ranging from very close to the ground to about 6,000 feet.

Stratocumulus consists of fairly large laminae, globular masses or rolls. They generally appear soft and grey, with denser parts. The elements of stratocumulus are arranged in a variety of ways, in lines or in waves. The rolls are often so close that their edges join, making a continuous layer. Layers of stratocumulus have a wavy appearance, sometimes exhibiting rather peculiar twists. There are transitional forms between stratocumulus and altocumulus and also between stratocumulus and stratus. Proper classification of these transitional formations is frequently very difficult.

Coronae are often observed in stratocumulus clouds. While the opinion is sometimes expressed that only the higher forms of stratocumulus show coronae, this does not appear to be confirmed by observation. Observation of such coronae is generally much easier by moonlight because the amount of general illumination is less. Sometimes the lack of uniformity in droplet sizes in stratocumulus clouds results in the appearance of an aureole rather than a corona. Irisations, which are common in altocumulus, are rather rare in stratocumulus.

Stratocumulus occurs fairly often below the middle cloud associated with overrunning warm air, both ahead of and during precipitation. Although stratocumulus is not usually a precipitating cloud, very light rain or snow occasionally fall from it. Showers of heavier precipitation falling from an overcast appearing to consist of stratocumulus only are traceable to cumulus congestus or cumulonimbus hidden from view by the stratocumulus. Sometimes these clouds of vertical development have bases below that of the stratocumulus, and they may then be seen to build up and pierce the layer cloud; at other times the bases of the stratocumulus and of the shower clouds are at the same level. In this latter case, the presence of cumulus congestus or cumulonimbus is revealed to the observer on the ground only by the occurrence of showers or of thunder or lightning. An observer above the layer cloud can see clouds of vertical development projecting above its upper surface.

Stratocumulus translucidus is a layer of cloud of no great thickness, and has interstices between the elements, through which blue sky is sometimes observed. When the sky cannot be seen, the thinner parts have a lighter appearance.

Stratocumulus opacus is a thick layer in the form of a continuous sheet of cloud with large dark rolls or rounded masses. The rolls stand out in relief from the under surface of the layer. This form of stratocumulus is a common winter cloud in middle latitudes.

Stratocumulus vesperalis is a flat elongated cloud which often forms about sunset, being the final product of the diurnal changes of cumulus. These clouds are formed when the tops of the cumuli dissolve, and the bases spread out. Sometimes, when convection is not very strong, this change takes place fairly early in the day. In such cases the term "vesperalis" seems to be scarcely appropriate.

Stratocumulus cumulogenitus is a fairly uniform moderately thick layer formed by the spreading out of the tops of cumuli. When the change takes the form of spreading out of the bases, as in formation of stratocumulus vesperalis, and occurs early in the day, "cumulogenitus" would seem to be appropriate.

Stratocumulus undulatus is a uniform layer of cloud with a roll structure. It was formerly known as "roll cumulus" in England and Germany. On account of the effect of perspective, this cloud form is very easily confused with flat cumulus clouds arranged in lines. When such cumuli, arranged in lines, merge, the product may take the form of stratocumulus undulatus.

Stratus is a generally uniform layer, and is usually low. In structure it resembles fog. Stratus may quite easily be confused with stratocumulus on the one hand and nimbostratus on the other. The lower surface has a very indefinite appearance, which makes estimation of height difficult. The cloud quite commonly envelops hilltops. Stratus is grey in colour, and is frequently accompanied by a layer of fog or haze between its base and the ground. The boundary between cloud and haze is not by any means easy to determine, either from the ground or from the air. The base has ragged projections hanging from it, but does not have the systematic arrangement of rolls or rounded elements displayed by stratocumulus. Stratus is not infrequently formed by the merging of fractostratus elements to form a layer of continuous cloud. The base of stratus displays on occasion lighter and darker spots, but regular features do not occur.

Stratus often imparts a hazy appearance to the sky. Confusion with nimbostratus need not occur if the cloud is producing precipitation, for stratus produces only drizzle or very light snow. If the cloud in question is producing real rain or snow it must be classified as nimbostratus.

Stratocumulus quite commonly changes into stratus, and stratus changes into stratocumulus. Stratocumulus is almost always higher than stratus. When a stratocumulus cloud becomes lower, and its elements appear very large and very soft, so that the regularly arranged globular masses and waves disappear, the cloud has become stratus. When the base of a stratus cloud begins to show a regular structure with real relief, the cloud has become stratocumulus.

Fractostratus is ragged irregular cloud, nearly at one level, broken up into shreds and uneven patches. It is very frequently referred to as "scud" and, on account of the lowness of its base, often appears to be moving more rapidly than it really is.

Fractocumulus has a ragged form somewhat similar to that of fractostratus, but is very different in origin and nature. It appears in very irregular forms, which usually develop into cumulus. In strong winds, fractocumulus clouds have a tendency to be elongated in a peculiar manner, often at angles near the vertical. Fractocumulus is sometimes the end product of cumulus, although in a large number of cases clouds of this type dissolve in an orderly fashion, retaining their characteristic appearance until they have completely disappeared.

Nimbostratus is a dark cloud, usually grey. Although dark, it has sometimes a curious appearance—it seems to be illuminated from within. It is usually amorphous and quite uniform, but the base is sometimes ragged. Continuous rain or snow often falls from it. The intensity of precipitation from nimbostratus may vary from light to heavy.

Nimbostratus is sometimes classified as a cloud of vertical development. There seems to be some justification for this, in view of the great thickness of nimbostratus and the fact that it is sometimes quite turbulent.

The criterion sometimes used to distinguish a cloud as nimbostratus is the actual production of precipitation. This criterion is not entirely adequate. The definition implies that nimbostratus may exist without precipitation occurring. A cloud which is really nimbostratus may fail to produce rain or snow which reaches the ground. Sometimes, precipitation does fall from the cloud, but is confined to virga. These virga impart to the cloud base a rather diffuse and indefinite appearance, making accurate determination of the lower limits of the cloud difficult or impossible.

Fractonimbus is a term which was sanctioned in 1933 by the International Committee. It is used to refer to the low ragged clouds of bad weather. The name has some arguments in its favour, the principal being that it is associated with rain. However, since it does not itself produce precipitation, but is caused by the precipitation, fractostratus appears to be a more satisfactory name for it.

CLOUDS OF VERTICAL DEVELOPMENT

Clouds of vertical development must be placed in a group by themselves. They cannot be assigned to any of the groups classified according to height, as they commonly extend through all of the levels assigned to the other groups.

Their bases occur at heights from 500 to 10,000 feet or higher and their tops at heights from 1,000 to 40,000 feet, or higher.

Cumulus clouds are generally thick, with vertical development. The base has a characteristic flat appearance, and the top is somewhat dome-shaped.

Cumulus humilis, commonly known as "fine weather cumulus" is a cloud with a flat base and a rounded top. This form of cumulus has moderate vertical development. The central portions are shadowed; the edges are much lighter and reflect considerable sunlight.

Cumulus congestus is a heavy and swelling cloud resulting from the continued growth and development of cumulus humilis. It builds upward and has a turbulent look, the tops often taking the forms of towers. But towers do not necessarily develop; instead the top may be one large heavy mass of cloud. The upper portions are sharply outlined, with a rather hard appearance. Sunlight reflected from the heads of these clouds produces a dazzling effect. Cumulus congestus may become a shower cloud if it extends high enough above the freezing level.

Cumulonimbus is heavy, with very considerable development. The tops are piled up in mountainous masses. The extreme upper parts are fibrous, and often spread out in an anvil. This is the thickest and most active of all clouds. Cumulonimbus forms as a result of continued upward development of cumulus congestus, and will occur only in very unstable air, the specific humidity of which is high. This cloud produces moderate to heavy showers and thunderstorms, and is the only cloud in which true hail is formed. The criterion which decides whether a cloud is cumulus congestus or cumulonimbus is that cumulonimbus exists only when the cloud has penetrated high enough above the freezing level that the cloud top contains ice crystals in considerable quantity. A cloud is correctly classified as cumulonimbus only when the head has become composed wholly or in part of ice crystals.

Cumulonimbus incus with its large anvil-shaped head composed of ice crystals is the best known and most easily recognized form. The anvils of old cumulonimbus incus clouds, when broken away by the wind, form tonitro-cirrus.

Cumulonimbus capillatus has a top which is a distinct cap of ice crystals, rather than an anvil. It is sometimes a transitional form between cumulus congestus and cumulonimbus incus.

Cumulonimbus arcus has a very ragged base, with a turbulent appearance. From below, it appears arched, very black, and ragged.

Cumulonimbus mammatus has a striking mammilated appearance on the base of the cloud, or on the underside of the anvil, or both. Large festoons or pouches hang from the underside of the cloud. A cumulonimbus of this type, because it has also an anvil head, is known as *cumulonimbus incus mammatus*.

Cumulonimbus calvus or "bald cumulonimbus" is the least easily recognized of the cumulonimbus forms. It has no anvil, and at first no cirriform parts are distinguishable. The tops, however, lose their clear-cut contours and become "softened." This form of cumulonimbus is frequently identified by the precipitation which falls from it.

SPECIAL CLOUD FORMS

There are several very interesting special cloud forms, which provide useful indications of the condition of the atmosphere and of movements in it.

The *castellated* forms of cloud derive their names from their characteristic structure, which bears a resemblance to the battlements of a castle. They indicate the presence of updrafts having cross sections which are small relative to their heights. Such cloud forms indicate instability at the levels at which they occur. Castellated clouds occur most frequently at the middle cloud level, where they take the form of *altocumulus castellatus*. When they occur at the low cloud level they take the form of *stratocumulus castellatus*.

The *mammilated* cloud forms occur only under conditions of considerable instability. They have a characteristic appearance, with large loops, pouches, or festoons hanging from their lower surfaces. These pouches sometimes look as if about to detach themselves. As mentioned above, this structure is found on the undersides of the anvils of some cumulonimbus clouds, such clouds then being known as *cumulonimbus incus mammatus*. A layer form of mammilated cloud is known as *stratocumulus mammatus*.

Lenticular clouds have a characteristic form with thin edges. When viewed close to the horizon, they resemble lenses and when viewed from below they appear as plates. They sometimes take rather large and heavy forms, of which the "Contessa del Vento" is a good example. These clouds are sometimes described as "spindle-shaped" and sometimes as resembling cuttle-fish bones. They are probably the clouds described in Hamlet as "very like a whale."

Lenticular clouds occur at all levels, from near the ground to high levels. They have commonly a nacreous white appearance, sometimes showing irisations. The forms are subject to rapid transformations, often being unrecognizable after a few minutes. The most common form of lenticular cloud is *altocumulus lenticularis*. One of the special forms of lenticular cloud is the *Moazagotl* cloud; some clouds of this type are quite heavy—occasionally heavy enough to be mistaken for storm clouds. They occur under föhn conditions.

Pileus is a cloud closely allied to the lenticular forms. It is a mantle of thin white cloud formed just above the top of a developing cumulus congestus. Sometimes the cumulus continues to develop and pierces the pileus, leaving the latter draped over the head of the cumulus in the form of a "cap," or about the lower portions of the towers of cumulus in the form of a "scarf" around the "shoulders" of the cloud. Pileus has sometimes been treated as a detail of the cumulus about which it forms, the cloud being referred to as "cumulus pileus." It is, however, a separate cloud of an essentially different type, and is better treated as such. Pileus has also been considered to be an aberrant form of *altocumulus translucidus.*

Banner clouds, which resemble flags flying from the tops of mountains, are caused by the presence of the mountains in the currents of air in which the clouds form. Such clouds sometimes serve to indicate hills which are not visible because they are below the horizon. This is most noticeable when the hill is on an island. *Crest* and *cap* clouds are formed also near and on hills, sometimes in contact with the hills, and sometimes separated from them.

The *Chinook Arch* is a characteristic cloud formation observed in the lee of the Rocky Mountains under Chinook conditions. It presents the appearance of an arch-shaped expanse of clear sky immediately above and to leeward of the mountains, with a fairly thick layer of cloud to leeward of the clear space. The arch form is apparent rather than real, being an effect of perspective. The associated cloud is a layer of altostratus having a thickness of several thousand feet, and a width varying from fifty to several hundred miles.

STRATOSPHERE CLOUDS

Certain rare cloud forms occur at very high levels in the atmosphere.

Noctilucent or *night luminous* clouds occur at heights of the order of 80 kilometres. They have a form very similar to that of cirrus, sometimes showing colours. On account of their great altitude they are illuminated by the sun long after it is below the horizon.

Mother of pearl or *nacreous* clouds occur at levels of 22 to 30 kilometres. They have forms resembling altocumulus, and show brilliant colours. These clouds also are illuminated by the sun when it is below the horizon.

BERGERON'S PROPOSED CLASSIFICATION

The system of cloud classification at present in use leaves much to be desired. Height is not the best basis, especially in view of the fact that the heights of the same cloud types vary so widely. A classification based on the actual nature of the clouds would have some definite advantages. A classification of this type was proposed by Bergeron in 1933. It is outlined below as an example of what the present writer considers would be a desirable trend in cloud classification.

Bergeron stated that a classification of clouds based on "an understanding of their physical and meteorological genetics and structure," and not on the outward appearance of the clouds only, would be desirable.

He suggested a classification based on the physical properties of the cloud elements and on their manner of formation. The proposed classification included six classes, of which three have solid elements and three have not.

1. *Simple crystals only.* Colloidally stable. The convective form would correspond to cirrocumulus, and the layer form to cirrostratus nebulosus.

2. *Simple crystals and hexagonal skeletons.* Colloidally slightly unstable. The convective form corresponds to cirrus unicinus, and the layer form to cirrostratus filosus or cirrocumulus.

3. *Hexagonal skeletons and fog droplets.* Colloidally absolutely unstable. The convective form corresponds to winter cumulonimbus and the layer form to altostratus.

4. *Fog droplets only.* Colloidally stable. The convective form corresponds to cumulus, and the layer form to stratus, stratocumulus or altocumulus.

5. *Drizzle droplets.* The corresponding convective form is cumulus and the layer form stratus, stratocumulus or altocumulus.

6. *Rain droplets.* The corresponding convective form is cumulonimbus and the layer form nimbostratus.

These may be subdivided according to whether the clouds result from rapid local ascent of air (convection) or slow ascent on a sliding surface.

2. Material of the Clouds

A CLOUD is a colloidal suspension of water in air and may be described as an *aerosol*. This was established by Wigand and Schmauss in 1929. The suspended water is in the form of liquid droplets or ice crystals or mixtures of droplets and crystals. Water droplets in clouds are often supercooled, and it is not unusual to find them in a liquid state at temperatures far below freezing point, even as low as $-50°C$.

Cirrus and cirrostratus clouds are composed of ice crystals, often in the form of needles or of hexagonal plates. Cirrocumulus, on the other hand, presents a rather more complicated picture. There is evidence that cirrocumulus sometimes is composed of supercooled water droplets, even at very low temperatures. This evidence of the presence of water droplets in cirrocumulus clouds consists of irisations and coronae, which are occasionally observed. One very fine example was observed by the writer at the airport at Ottawa, Ontario, at 1440 Eastern Daylight Saving Time on February 8, 1944. Beautiful coloured irisations were observed near the sun in cirrocumulus at an estimated altitude of 20,000 feet. The colour of the irisations was predominantly a slightly brownish red, rather like autumn leaves. The crest of a ridge of high pressure had just passed, and cirrus and cirrocumulus were in the sky. It is reasonable to suppose that at the level of the clouds there was some advection of returning modified polar air, flowing up the slope of the dome of fresh cold air and expanding with its ascent to cause formation of high cloud.

The water found in middle clouds is subject to rather more variation of phase than in high clouds. High clouds are generally well above freezing level; while middle clouds in temperate latitudes are found quite near freezing level in summer, and above it in winter. Altocumulus and altostratus are composed sometimes of ice crystals and sometimes of water droplets. The thinner forms of altostratus are usually composed of ice crystals. This has been quite adequately demonstrated by the presence of halos and parhelia in clouds which are of a grey colour, and should therefore not be called cirrostratus. Altostratus and altocumulus, however, are most commonly composed of water droplets, which are often in a supercooled condition. These give rise, in altocumulus, to irisations. The thicker forms of altostratus are, in most cases, composed of ice crystals in their upper portions and of water droplets in their lower portions. These statements apply to middle latitudes; the higher freezing levels prevailing in warmer latitudes will result in a much larger proportion of the middle clouds being composed of water at temperatures above freezing. This is to some extent counteracted by the tendency of the middle clouds to occur at higher altitudes in warmer latitudes.

The low clouds are usually composed of water droplets, sometimes at temperatures above freezing and sometimes supercooled. However, at extremely low temperatures, low clouds are often composd entirely of ice crystals. Fractostratus of quite ordinary form and appearance has been observed in very cold weather to display optical phenomena peculiar to ice crystal clouds. For example, the writer has observed a parhelion in fractostratus based at approximately 1,500 feet.

The forms of water in nimbostratus clouds display considerable variety. In the warmer seasons of the year, and in warmer air masses in colder seasons, there is a region near the base of a nimbostratus cloud composed of fairly large drops of water. Higher in the cloud will be found a region where supercooled droplets are present; then higher again a region of mixed droplets and ice crystals and finally, in the thicker nimbostratus, a region composed of ice crystals only. The upper portion of nimbostratus generally merges with the altostratus which precedes it. In cold weather, there is snow rather than water drops near the base of nimbostratus. The different constituents of the cloud are not sharply divided into "regions" with well-defined boundaries; there is a more or less gradual transition.

Stratus clouds are generally composed of water droplets, sometimes supercooled. Occasionally at low temperatures stratus is composed of ice crystals. However, unless they definitely have the form of stratus, such clouds are in most cases better classified as cirrostratus. Fractostratus, although occasionally composed of ice crystals at very low temperatures, is most frequently a water droplet cloud. Mixtures of ice and water are not to be expected in stratus or fractostratus. Thickness of these clouds is too small to permit of appreciable temperature variation within them.

Stratocumulus is usually made up of water droplets, except on rare occasions at low temperatures, when it is composed of ice crystals sometimes merged to form snowflakes. The water droplets are often supercooled in colder seasons. Mixtures of water and ice are not usually found in stratocumulus clouds, but considerable variety in the size of the water droplets within the same cloud is to be expected.

The composition of clouds of vertical development is interesting. Cumulus humilis is composed of water droplets, usually at temperatures above freezing; except that supercooled droplets occur on the rare appearances of this cloud in winter. When cumulus congestus develops, it often builds up to above the freezing level, with the result that the upper portions are then composed of supercooled droplets. Additional development to heights far above the freezing level results in the formation of ice crystals in the tops. When this occurs, the cloud becomes cumulonimbus. The presence of ice crystals in the top of cumulonimbus is highly significant, on account of the very close relation of ice crystals to the release of precipitation. Cumulonimbus contains a notable variety of components. The lower portions are composed of water drops, those above the freezing level being supercooled. The highest part of a cumulonimbus is made up of ice crystals, with no water droplets. In some of the most active clouds of this type, hailstones in various stages of development occur.

The composition of the very high cloud forms of the stratosphere is not yet fully understood. Irisations observed in mother-of-pearl clouds indicate the presence of water droplets. Dust from various sources has been suggested as the material of the very high clouds. The sources suggested include dust from volcanic eruptions and from cosmic sources. The latter could be dust from meteors, from some of the planets, or from cosmic material given off by the sun.

The droplets in clouds were studied by Bricard on the Puy de Dôme, and later on the Pic-du-Midi. The result of this work established that clouds are not homogeneous with regard to droplet size. The lack of homogeneity is also indicated by the fuzzy aureoles sometimes observed instead of coronae. The radii of the droplets were found to vary in the ratio of 1 to 10 within the same cloud. However, for each cloud there was found to be a "favoured" size, the majority of the droplets being close to this size.

The favoured droplet size is characteristic of the cloud type concerned. In the centres of well-established clouds of various types the favoured radii were found to be: in nimbostratus, 11 microns; in stratocumulus, 8 microns; in cumulus, 6.5 microns; in stratus, 5.3 microns. In clouds in process of formation or of dissipation the droplets were found to be smaller than in well-established clouds. In the same cloud the favoured size is larger in the centre than at the edges, but is always of the order of a few microns.

Subsequent investigations performed with the aid of aircraft have produced similar results with respect to droplet sizes. Researches have been performed by the National Advisory Committee for Aeronautics and by the Mount Washington Observatory in the United States, by Mazur in England, and by Diem in Germany. The techniques employed have included catching cloud droplets on microscope slides; and collecting ice on rotating cylinders during flight in supercooled clouds.

On account of the much greater accessibility of fogs than of clouds most of the studies on droplet size have been made of fog droplets rather than of cloud droplets. In this connection, it is well to remember that applying the results to clouds introduces a more or less hazardous extrapolation. For example, studies of electrical charge made on fogs would not be a safe foundation on which to base conclusions regarding clouds, since the relative positions of clouds and fogs with respect to the earth are radically different.

Investigations of cloud conducted on mountain tops have disadvantages introduced by the motion of the clouds relative to the observer; moving objects are never as easy to study as those at rest. The use of aeroplanes appears to be objectionable for the same reason, and in this case the relative motion is considerably more rapid. It seems to the writer that the ideal arrangement would be to fly into the cloud in a dirigible and then drift with the cloud, with the engines stopped. Relative motion would thus be eliminated, and the effect of engine exhaust and of the turbulence of propellors removed. The development of devices to measure droplets from aircraft in such a way as to cancel the effects of relative speed would still not be entirely satisfactory. The passage of the aircraft, so long as it is moving relative to the cloud, must inevitably set up sufficient turbulence to disrupt the normal arrangement of cloud droplets or crystals in its vicinity and possibly also to cause a certain amount of artificially induced coalescence. Further, so long as engines are operating, conditions in the vicinity of the aircraft must be abnormal. This abnormality is the result of the appreciable addition of water vapour in the engine exhausts and also of the presence of a certain amount of carbon in the exhausts.

3. Processes of Cloud Formation

FORMATION of a cloud involves the transformation of some of the water vapour in the atmosphere to the liquid or solid state. This transformation may take the form of a change from vapour to liquid droplets, namely *condensation,* or a change from vapour to ice crystals, namely *sublimation.*

In order that clouds may form, certain conditions must be fulfilled. A sufficient number of suitable nuclei, on which cloud components may form by condensation or sublimation, must be present in the air. In nature, lack of such nuclei does not seem to be a serious factor. It is also necessary for the air to be in saturated or very nearly saturated condition. Continuous condensation requires some means of continuously supplying water vapour and so maintaining saturation.

The requisite condition of saturation may be brought about in various ways. These may be roughly classed as processes by which the air is cooled, with a consequent reduction in its capacity for water vapour, and processes by which water vapour is added to the air, with a consequent increase of the water vapour content to the maximum possible at the prevailing temperature. These processes may, of course, act together to bring about saturation.

The most common of the processes by which air is cooled, with resulting cloud formation, is that of expansion. This expansion is the result of lifting of the air to levels where the pressure is lower. The lifting may be caused by thermal updrafts, or by mechanical updrafts, or by a combination of these. Indirect lifting caused by ascent of underlying air also occurs, that is, a horizontal flow of air may be deflected upward by an updraft in underlying air.

Mixing of the lowest layers of the atmosphere often produces low cloud. Just to what extent mixing operates at higher levels is still largely in doubt. Clouds are sometimes formed indirectly. When air is cooled by contact with a cold land or water surface fog may form, and this fog may later be lifted to form cloud.

Throughout the present discussion on formation of clouds the assumption will be made that an adequate supply of nuclei is available. Nuclei of condensation and sublimation are discussed in the following chapter.

THERMAL LIFTING

Thermal lifting of air results from uneven heating of the earth by solar radiation. The air above the warmer spots becomes warmer and therefore lighter than the surrounding air, with the result that instability develops and updrafts are started. Since the air cannot all rise there must be compensating downdrafts, and a rather irregular pattern of updrafts and downdrafts is produced.

On account of the decrease of pressure with increased height in the atmosphere, rising air will expand and cool. The rate of decrease of temperature of air in such rising currents is normally the dry adiabatic lapse rate. If the updraft continues high enough, the air will become saturated and further lifting will nearly always result in condensation of some of the water. The latent heat released in this process reduces the rate of cooling of the rising air to the saturated adiabatic lapse rate. The level at which condensation begins is known as the *convective condensation level*. The first condensation to occur is normally in the form of extremely small water droplets, which constitute merely a haze layer. Since the air in such updrafts is usually rising quite rapidly, this haze layer is very shallow and the condensation products soon take the form of cloud droplets. The shallowness of the haze layer accounts for the clear-cut appearance of the bases of cumulus clouds.

In flight over mountainous terrain, the writer has observed the interesting phenomenon of cumulus clouds forming over certain valleys at lower levels than over most of the rest of the terrain. This would appear to indicate a somewhat higher moisture content in the air rising from these valleys, and hence a lower convective condensation level than in most of the air mass.

The vertical extent of clouds formed by thermal lifting is determined by the moisture content and stability of the lifted air. In air containing a large amount of water vapour, more water is available and so a thicker cloud may be produced. In unstable air, the vertical currents extend to greater heights, tending to thicken the clouds. The amount of heating on the particular occasion and the nature and effectiveness of other cloud-forming agencies present also profoundly affect the development of clouds formed by thermal lifting.

The first cloud produced by thermal updrafts usually takes the form of fractocumulus. Fractocumulus is a ragged cloud, and takes very irregular forms, sometimes considerably elongated in a vertical direction, the cloud being inclined at an angle to the vertical. This is the natural result of the downwind displacement of thermal updrafts, a phenomenon well known to sailplane pilots. Veils of mist observed to precede the actual formation of these clouds are sometimes known as *fumulus*.

Sufficient heating and an adequate supply of moisture result in continued development of fractocumulus to form cumulus humilis. This cloud forms usually in a layer of air where the convective condensation level is not far below a layer of stable air. The updrafts generally do not penetrate to any appreciable extent into the stable layer, as their energy is consumed in that layer. In other words, cumulus humilis forms and remains as cumulus humilis when the convective condensation level is not far below the top of the rising currents of air.

Very frequently when cumulus humilis forms during the afternoon it dissipates after the time of maximum incoming solar radiation has passed. In such cases, the base of the cumulus often becomes progressively higher as the afternoon goes on and the heating of the air raises the convective condensation level. The clouds rise in their entirety, generally becoming smaller, and retaining their characteristic form. When the earth begins to cool on account of outgoing radiation, there is a tendency for the cumulus to sink. This sinking causes compression and heating, with consequent dissolution of the clouds.

Cumulus humilis may form stratocumulus cumulogenitus by increasing in amount until a complete layer of cloud is formed by the merging of the cumuli. In the late afternoon, a layer or patches of stratocumulus vesperalis may be formed by spreading of the bases of cumulus clouds, while the tops dissolve. The tops of cumulus clouds may spread out while the bases dissipate, so forming altocumulus cumulogenitus, in a layer or patches.

When cumulus humilis continues to develop, formation of cumulus congestus results. This occurs only when there is a deep layer of conditionally unstable air, containing sufficient water vapour to provide droplets in great enough quantity to form a large cloud. In these cases considerable energy is released, and the cloud tops develop rapidly, often taking the form of towers.

There is considerable turbulence in cumulus congestus. Sometimes the vertical currents are sufficiently strong and extend sufficiently high that the cloud penetrates above the freezing level. The tops are composed of water droplets and have a hard appearance, somewhat resembling a cauliflower. A very active aspect often appears to "boil."

Continued growth and development of cumulus congestus gives rise to formation of cumulonimbus. The transition shows clearly in the structure of the top of the cloud. The hard outlines soften and ice crystals often extend out from the top in streamers.

Development of cumulonimbus involves a considerable mixture of ice crystals in the top of the cloud. This requires penetration to well above the freezing level. Above the $-10°C.$ isotherm ice crystals predominate strongly, and in the extreme tops of cumulonimbus clouds ice crystals only will be found. Below the $-10°C.$ isotherm supercooled water droplets are found in considerable numbers.

The streamers of ice crystals at the top of a cumulonimbus sometimes spread to develop into an anvil-shaped head, forming the well-known cumulonimbus incus. The spreading out of the anvil appears to be caused in part by a stable layer of air. This stable layer tends to arrest vertical development, because the energy of the updrafts in the cloud is consumed by the resistance to vertical motion which is always offered by stable air. Sometimes, when the wind at the level of the tops of cumulus congestus is strong and the moisture content of the air is not very great, the towers are torn off by the wind and the cloud droplets evaporated into the air above. In such cases, the cloud top sometimes presents an appearance deceptively like that of the top of a cumulus congestus which is in process of becoming cumulonimbus.

The change in the top of cumulus congestus when it becomes cumulonimbus may be relatively inconspicuous. The hard outlines are softened and a few streamers of ice crystals appear. The cumulonimbus so produced is cumulonimbus calvus, known also as "bald cumulonimbus." A cloud known as cumulonimbus capillatus is intermediate between the "calvus" form and the "incus" form.

A rare type of cumulus is formed by forest or bush fires. The considerable heat generated by such a fire causes a powerful updraft, and water vapour in considerable quantities is liberated by the burning wood. This updraft with the addition to the air of water vapour is occasionally sufficient to cause formation of a cumulus cloud.

Cumulus clouds sometimes occur in quite straight lines. Such arrangements of cumuli are known to aircraft pilots as *cloud streets*. These rows of clouds have been attributed to the presence of longitudinal eddies in a conditionally unstable layer of air, the cumuli forming in the ascending branch of the circulation.

FRONTAL LIFTING

Extensive cloud systems are produced by the ascent of air resulting from interaction of air masses at fronts. The character and extent of these frontal cloud systems are determined largely by the type of front and its slope and motion. The moisture content and stability of the air masses concerned are also important in determining the nature of the frontal clouds.

Formation of clouds at fronts depends entirely upon the presence of sufficient water vapour in the warm air for saturation to occur when the air is lifted and cooled. When the warm air is very dry, no frontal clouds form. Topographical features affect frontal clouds in varying degrees.

Warm front cloud systems are caused by a fairly gentle flow of warm air over a wedge of cold air. As a warm front approaches, the first and highest cloud observed is cirrus. As the front moves towards the point of observation, the layer of air affected by the lifting becomes deeper, and the cirrus thickens to cirrostratus. Thickening continues, with transformation of the cirrostratus to thin altostratus. Further thickening produces thicker and darker altostratus, which becomes altostratus precipitans. From this cloud precipitation usually commences. In many cases, altocumulus is mixed with the altostratus. The cloud continues to thicken, and becomes nimbostratus, which is the principal precipitating cloud.

Rain falling from the base of altostratus and nimbostratus has a profound effect on the air below the cloud. Very often, rain evaporates, taking its latent heat of vapourization from the air below the frontal surface. This often causes slight supersaturation of the air below the cloud, followed by condensation of the surplus water to form additional cloud or fog. The moisture content of the air, and the strength of the wind will determine whether the result will be cloud or fog. Supersaturation with consequent formation of cloud is usually effective immediately below the base of the nimbostratus, causing it to build downward below the front. At other times, supersaturation and consequent cloud formation occur at lower levels, forming very low fractostratus separated from

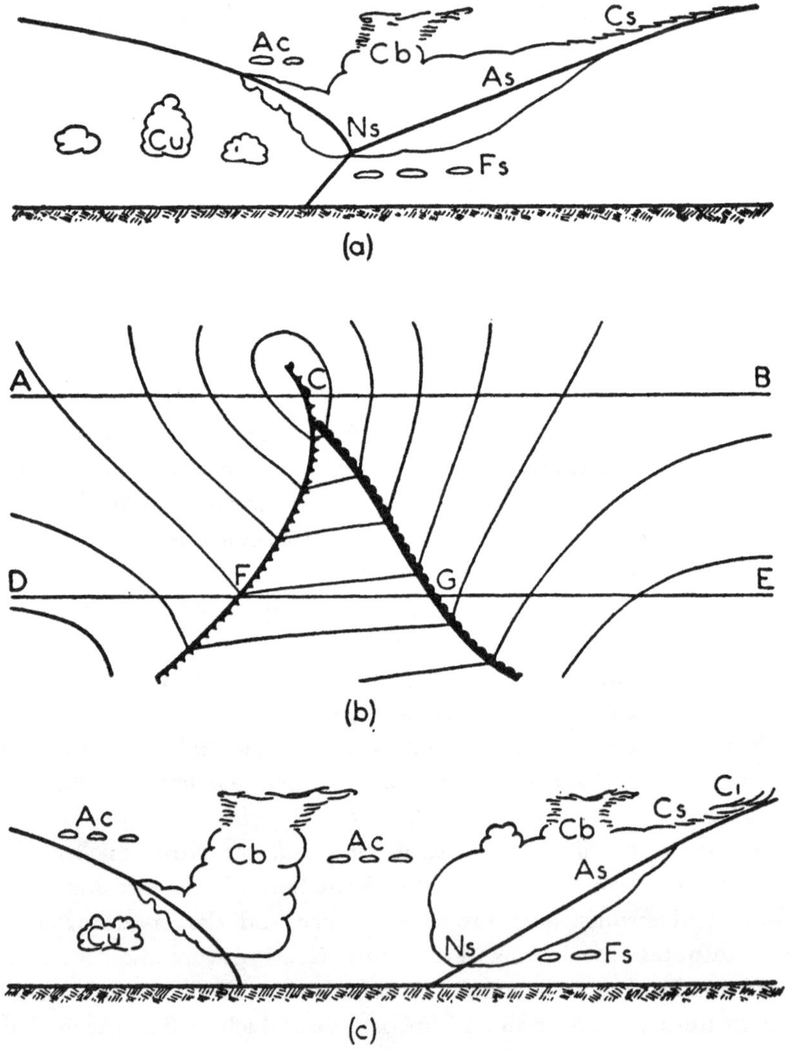

Fig. 1. Typical clouds in a frontal depression of the type shown in (b); (a) cross section through the occlusion, (c) cross section through the warm sector. (Reproduced by permission from *Meteorology Theoretical and Applied*, by E. W. Hewson and R. W. Longley, published by John Wiley & Sons, Inc., 1944.)

the precipitating cloud. Sometimes, the fractostratus merges to form a layer of stratus. Mixing in the lowermost layers of air is an important factor in the formation of this low fractostratus.

If the moisture content of the overrunning air is very high and the air is unstable, cumulonimbus clouds will form. The cumulonimbus so formed has its lower parts in the altostratus or nimbostratus, and the turbulence does not extend below the frontal surface.

The warm air is lifted more violently at a *cold front* than it is at a warm front, and more active types of cloud are formed. The more rapid lifting leads to formation of cumulonimbus clouds if the air of the warm sector is sufficiently unstable and contains sufficient water vapour. The clouds of vertical development along a cold front are generally preceded by a layer of altocumulus. Sometimes, there is considerable subsidence in the warm air above the frontal surface. This subsidence produces inversions, which may be sufficiently strong to impose limits on the vertical development of the cumuliform clouds at the front. These limits may be so severe as to prevent entirely the development of cumuliform clouds.

When the warm air becomes occluded, the cloud systems resemble frontal cloud systems. With both warm and cold types of occlusion, cloud systems are found which resemble both regular types of frontal cloud systems. It will generally be found that the clouds at a cold front occlusion are of a more active type than those at a warm front occlusion.

In numerous situations a transition from cold to warm air occurs, but the transition is not sufficiently sharp to constitute a front. This is common on the rear of high pressure areas, where zones of convergence frequently occur. It is particularly notable after the crest of the ridge of high pressure has passed. The warmer air is forced to rise with consequent cooling and cloud formation much like that caused by a warm front. Extensive cirrus in high pressure areas is often formed in this manner. If the moisture content of the air is great enough, precipitation will occur. When the moisture content is great and there is a high degree of instability, convergence results in formation of cumulonimbus clouds.

OROGRAPHIC LIFTING

Topographic features play a considerable part in cloud formation processes. The ascent of air caused by the presence of a hill in the air stream frequently causes sufficient expansion and cooling to form cloud. The extent to which the air is lifted is influenced by the strength of the wind, the height and steepness of the hill, and the stability of the air. Orographic clouds vary from the most insignificant fractostratus or fractocumulus to very large and active cumulonimbus and nimbostratus. The moisture content of the lifted air governs to a very considerable extent the amount and type of cloud formed.

The relative locations of orographic clouds and hills which cause their formation are subject to considerable variation. The cloud may rest on the hill, above it or to leeward of it, or may form a ring about it.

In stable air the relative humidity of which is high, stratus cloud is often formed, covering hilltops. When the air is stable and the wind moderate, there is an orderly flow of air over the top of the hill and down the lee side. Cloud so formed has a well-defined base on each side of the hill. The base on the windward side is at the level where saturation occurs as a result of the lifting, that is, at the *lifting condensation level*. On the lee side, the cloud base is at the level where the cloud has been completely evaporated by the heat resulting from compression of the descending air. If no precipitation occurs on the windward side, the air loses no water, and the same amount of heat as was given up by the condensing water on the windward side will be required to evaporate the cloud on the lee side. Consequently, in this case, the cloud base on the lee side will be at the same level as the cloud base on the windward side. The term "crest cloud" is sometimes applied to this form of stratus.

When air containing a large amount of water vapour flows strongly against a sufficiently high hill, the orographic cloud thickens to form nimbostratus. Precipitation in considerable quantities then falls on the windward slope. Since water is lost by precipitation, the cloud base on the lee side of the hill will be higher than that on the windward side, because the smaller amount of water remaining in the air is evaporated with less heat than was given up by the condensing water on the windward side. The air then continues to flow down the hill, being warmed by compression, and arrives at the base of the hill as a warm dry wind. In this manner the Chinook winds of the Rocky Mountains and the föhn winds of Europe are produced. In India, very heavy rain is produced in this manner by air flowing against the Western Ghats, and a hot dry wind blows into the plain beyond.

Cumulus may be formed by orographic updrafts, and lines of cumuli are often observed above hills on days when no cumulus is forming elsewhere. These cumuli are not necessarily in contact with the tops of the hills, because air, on striking hills, is sometimes forced to rise to heights considerably above their tops. Just how far it must rise to form cumulus clouds depends upon the convective condensation level, which is, in turn, determined by the moisture content of the air.

If the air is sufficiently unstable, and contains a sufficient amount of water vapour, development of cumulus cloud formed by hills may continue until the cloud has become cumulonimbus. The bulk of such cumulonimbus will be on the windward side of the hill or hills.

Certain special cloud forms occur only as a result of the presence of hills in the current of air. As already pointed out, the positions of such clouds relative to the hill or hills concerned are not always the same.

Cap clouds form above mountain tops, taking the form of the mountains without actually touching them. Such clouds are a form of cumulus developed from ascent of air up the slope. The cap does not show much lack of symmetry, although the ascent of the air is naturally rather asymmetrical. The cap forms continuously on the windward side, and dissolves continuously on the lee side of the hill.

Sometimes, in strong winds, the ascending air reaches its maximum elevation on the lee side of the hill, and a roll of cloud is formed. While this roll bears some resemblance to the lenticular clouds, it does not have thin edges.

On the lee side of a hill, the orographic clouds usually dissipate, and in dissipating take a wide variety of forms. One form, associated with föhn winds and

FIG. 2. Positions of orographic clouds relative to the hills which cause their formation. (From *La Physique des Nuages*, by J. Coulomb and J. Loisel, Editions Albin Michel, Paris, 1939).

commonly referred to as the föhn wall, is a rather massive cloud, having the appearance of a wall when viewed from below.

In some instances, clouds descend the lee side of a hill like a cataract. At other times, a banner of cloud resembling the smoke from a locomotive will persist. Banners, however, are not necessarily always the remains of cloud formed on the windward side. They may be newly formed, the air having been lifted just high enough for cloud formation to occur. A theory advanced by Humphreys is that such banners are formed in the "wake" of the hill as a result of expansion and cooling in a small region of low pressure.

An interesting example of banner cloud is that occurring on the Matterhorn, giving it the name "Matterhorn which smokes." A study of this particular cloud was made by Kampé de Feriet, with the aid of a motion picture camera. The results indicated the possible existence of two eddies with horizontal axes on the lee side of the mountain. Such eddies provide a useful explanation for the presence of "stationary" clouds on the lee side of a mountain. These rather rare clouds are formed in the ascending branches of such eddies, and dissipated on their lee sides. Thus, while the matter of the cloud is constantly changing, the cloud itself is effectively stationary. Such large eddies appear able to exist only with wind speeds which do not exceed 10 metres per second, or 22 m.p.h. The exact limiting speed varies with the stability of the air. If the wind speed is too great, the eddies are replaced by a turbulent, disorderly motion. If the wind is too light, or if the air is too dry for a sizeable cloud to form, the banner is reduced to a light "smoke." Tenuous streamers of this kind have sometimes caused inexperienced observers to report "smoke" issuing from extinct volcanoes.

INDIRECT LIFTING

Clouds do not invariably form in the layers of air directly disturbed. Sometimes they form in layers of non-turbulent air disturbed by the motions of air below. This will occur only when the relative humidity of the non-turbulent layer is high, because there is not sufficient vertical motion to cause a large temperature drop.

Lenticular clouds are formed by the lifting of damp, non-turbulent layers of air. The lifting originates from upward motion imparted to the air by some *obstacle*. The obstacle may be a hill or hills. In this case, motion of the lower layers of air is communicated to layers with high relative humidity above, the lifting being sufficient to cause cloud formation. In order that this may occur, there must be present at higher levels stratified layers of air with a sufficiently high relative humidity. In the simplest case, this lifting consists of raising, from a position of rest, a "dome" of air. Lenticular clouds occur at a wide variety of heights, but are most prevalent at middle cloud levels. They occasionally assume rather large and heavy forms, of which the "Contessa del Vento" is an example.

The obstacle need not be a visible one such as a hill. Sometimes, an invisible obstacle in the form of a convective element supplies the disturbance. The convective element concerned in this process is a column of rising heated air, which need not necessarily have any condensation within itself.

If there is an inversion above the damp layer, it offers resistance to convection from below, with the result that vertical motion is slight, and lenticular cloud forms. If, on the other hand, convection from below pierces the inversion, fractocumulus or one of the castellated forms of cloud will result. Letzmann has reported observing castellated forms of cloud pushing against the dome of lenticular clouds. He explains the lenticular cloud as caused by lifting provided by a "last push," this push being provided by energy made available by the liberation of the latent heat of vapourization, and being sufficient to break through the inversion.

Irregularities in the terrain initiate motions in the atmosphere which sometimes produce rather complicated eddies. Such eddies may cause a series of waves, which waves may then transmit disturbances to higher layers of air of high relative humidity and so cause formation of a succession of lenticular clouds in single file. The best examples of such series of waves are found on the lee sides of mountain chains. The first part of a wave to occur downwind from the hills may be an updraft; under other conditions, there is an initial sinking. In order that cloud may form, the amplitude of the waves must be great enough to provide a lift sufficient to bring about saturation in the damper layers above.

An outstanding example of cloud formed by such a series of waves is the cloud known as *Moazagotl*. This type of cloud occurs generally in föhn seasons, and at quite high levels.

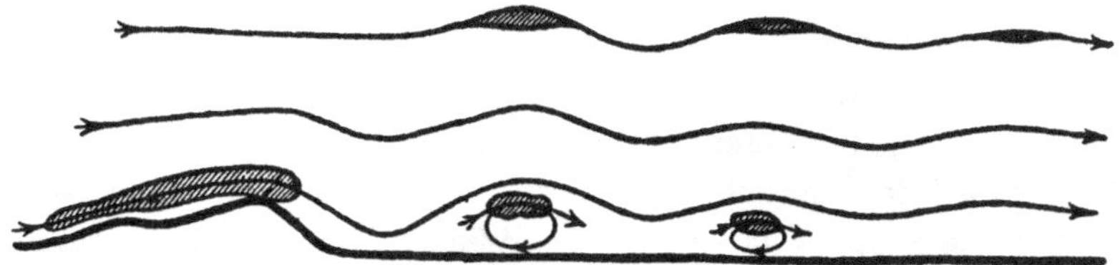

Fig. 3. The Moazagotl type of clouds and the system of standing waves which causes them. (From *La Physique des Nuages*, by J. Coulomb and J. Loisel, Editions Albin Michel, Paris, 1939.)

The eddy systems which cause these clouds were studied in detail by Küttner. His work is particularly interesting, in view of the fact that he used a sailplane to investigate the air currents. No turbulence was observed at the base of the region where the waves were produced. Lenticular clouds were present and also a mass of cloud on the windward side of the mountain. In addition there were rolls of cloud in the form of elongated bands produced by turbulent eddies with horizontal axes. One of these *rotors* was found under each of the waves producing the lenticular clouds. Küttner reports very violent turbulence in the rotors. The stratified layers, however, were found free of turbulence.

A species of lenticular cloud formed notably in the vicinity of cumulonimbus is attributed to the meeting of a dry descending current of air and a moist ascending current. Letzmann ascribed the formation of lenticular clouds in general to this process. The theory is illustrated in the accompanying diagram.

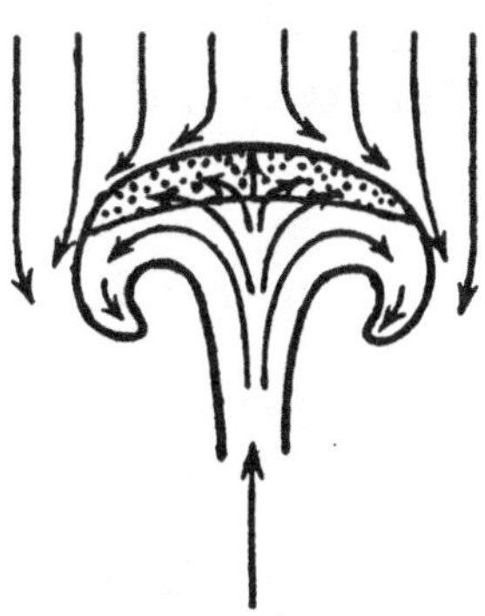

Fig. 4. A lenticular element, and the air currents which Letzmann's theory postulates as responsible for its formation. (From *La Physique des Nuages*, by J. Coulomb and J. Loisel, Editions Albin Michel, Paris, 1939.)

Formation of lenticular clouds was considered by Letzmann to be caused by rising columns of damp air, like those producing cumulus clouds. These rising columns were assumed to meet descending currents of cold air, with the result that clouds formed in the rising air were flattened and smoothed down. This theory is radically different from the more generally accepted theory that there are multiple surfaces of inversion caused by subsidence. Lenticular clouds do occur under föhn conditions with descending currents, but they form at an elevation far above that of the descending currents in the lee of the mountains.

The *pileus* is sometimes treated as a detail of cumulus, but is really a separate cloud. It is closely related to the lenticular clouds, being formed by a process which is very nearly the same. Rising air pushed up by developing cumulus heads produces a local dome-shaped deformation in the flow of the layers of air above, and if these upper layers have a sufficiently high relative humidity, a rather delicate white cloud forms. The pileus forms above, and separated from, the head of the cumulus, but continued vertical development of the cumulus frequently causes it to overtake and pierce the pileus. Some very interesting effects are produced as a result. When the cumulus overtakes the pileus, a "scarf-like" effect is produced. Such a pileus is known as a *scarf cloud*. On other occasions, when the cumulus pierces the pileus, descending currents in the vicinity of the sides of the cumulus tend to cause an annular dissolution of the pileus. This annular dissolution sometimes continues until the pileus is reduced to small bars of grey cloud, entirely detached from the cumulus which produced it.

Pileus clouds sometimes have a layer or leaf construction, indicating a series of surfaces of inversion. Wegener has observed correspondence between four lenticular clouds and four inversions, demonstrated by soundings.

The cloud associated with the *Chinook Arch* appears to be formed as a result of indirect lifting, but the exact nature of its formation has not been explained. A warm layer of air apparently crosses the Rocky Mountains at a high level. Air at a lower level, deflected upward by the mountains, appears to raise the warm layer enough to cause condensation and thus form the typical Chinook layer of altostratus. The opinion has been expressed that the cloud is of the Moazagotl type.

CONTACT COOLING

Contact cooling may indirectly give rise to the formation of cloud. In the first instance fog is produced, and this in turn is lifted by thermal or mechanical updrafts to form cloud.

Advection fog may be lifted to form low stratus, which is sometimes quite extensive. Sea fog, forced by a fairly strong wind over a coastline or island, may be lifted to form cloud. The height of its base is determined by the nature of the terrain and the strength of the wind. Advection fog formed inland may produce cloud in the same way. Whether the stratus formed in this fashion remains cloud or reverts to fog is determined by the strength of the wind and the nature of the terrain downwind from the original lifting features. Should the wind be light, or the terrain too flat to produce sufficient disturbance, the fog remains fog. Heating from below occurs if the fog is formed in the daytime, and this may produce updrafts sufficiently strong to dissipate the stratus. This process will be assisted by a relatively small amount of solar heat absorbed by the cloud's upper surface. If the air is very stable and has a high specific humidity, heating will often serve to thicken the cloud.

When radiation fog begins to break up under the influence of solar heating or of wind or both it quite often lifts to form low fractostratus as it dissipates. In such cases the low cloud is generally of short duration.

Formation in this manner of very short-lived fractostratus was observed and photographed by the writer in the valley of the Gatineau River, near Ottawa. (See Plate XXV). In the Gatineau River valley radiation fog is common, especially in the autumn, because there is considerable drainage of cold air into the valley. The fog, on the occasion illustrated, was confined to the river valley, being topped 700 to 800 feet above the water level. Shortly after sunrise, the fog lifted from the water, becoming very low fractostratus cloud, which dissipated as it rose. When the fractostratus had risen high enough to be noticeably affected by the wind it became torn and very ragged. At about 500 feet above the water the cloud was completely dissipated. By the time the lifting began, the fog had thinned to less than its original depth, becoming very shallow. The fractostratus was evidently evaporating as convection currents carried it to drier air a short distance above the surface. When fog is deeper and more extensive, the stratus or fractostratus formed from it lasts longer.

MIXING

Cloud is frequently formed as a result of mixing of the lowermost layers of the atmosphere. When the surface wind is fairly strong, turbulence is set up, with the result that the air near the ground is thoroughly mixed. Von Bezold showed that mixing between two "parcels" of air of different temperatures may lead to saturation or even to supersaturation, although neither of the "parcels" of air was necessarily saturated before the mixing took place.

Petterssen showed that *horizontal mixing* across a zone of transition, even with a large initial temperature difference, and high initial relative humidity, would lead to only a very small amount of condensation. From this the conclusion is drawn that horizontal mixing is negligible as a fog-producing process. It is further pointed out that for initial relative humidities of less than 98 per cent, saturation would not be reached in this way.

Vertical mixing results in the establishment of an adiabatic lapse rate and constant specific humidity in the mixed layer. If there are no changes in moisture and heat content the mean temperature and mean specific humidity are constant.

Vertical mixing tends also to decrease the relative humidity in the lower portions of the mixed layer, and to increase the relative humidity in its upper portions. This makes it clear that vertical mixing is not a producer of fog, but is a prolific producer of cloud.

The occurrence or non-occurrence of condensation in the upper portion of the mixed layer will be determined by three factors: the amount of water vapour in the layer, the vertical distribution of temperature and the thickness of the layer. As the depth of the mixed layer increases, the amount of cooling increases. An increase in the amount of cooling increases the chance of condensation in the top part of the layer. The height at which saturation is attained is the *mixing condensation level.* If an inversion is present at a fairly low altitude, and the wind is reasonably strong, it may be assumed that the air below the inversion is completely mixed. It has been demonstrated by Petterssen that the amount of condensed water produced as a result of vertical mixing is of the correct order of magnitude to form cloud.

The thickness and relative humidity of the mixed layer will govern the height of the base of the cloud formed. When it is a shallow or very damp layer, the result is low fractostratus or stratus. Deeper layers below the mixing condensation level result in formation of higher stratus or stratocumulus.

Formation of low stratus by mixing occurs typically when radiational cooling has caused considerable temperature reduction and consequent increase in relative humidity of the air nearest the ground. This condition must be accompanied by a strong enough wind to mix the lowest layer of air. At the Ottawa airport the author has observed that a surface wind of about 10 to 15 miles per hour is sufficiently strong. The most favourable time of day for mixing to occur and produce cloud is shortly after sunrise, when the sun has caused sufficient heating to impart to the air a vertical motion. This vertical motion assists the wind in mixing the lowermost layers of air. In practice, it has been found that a fairly accurate estimate may be made of the height at which stratus will be formed under these conditions. If a dry adiabatic lapse rate is assumed, and the surface temperature and dew point are considered, an approximation of the height of the mixing condensation level may be easily obtained. For example, stratus at 400 feet has been successfully predicted with a surface temperature and dew point spread of two Fahrenheit degrees, and a surface wind of about 15 miles per hour.

As the sun rises higher, solar heating frequently warms the air sufficiently to cause the cloud to evaporate. This process is particularly effective when the cloud is very thin. However, if the relative humidity of the air is high for a considerable distance above the surface, additional heat from the sun causes more mixing and thickens the layer of stratus so much that it sometimes persists for the entire day. This condition is likely to prevail in a warm sector. When stratus is produced by mixing at night it is to be expected that radiational cooling of the cloud's upper surface will result in thickening of the cloud.

ADDITION OF WATER VAPOUR

Cloud may be formed as a result of the addition of water vapour to the air, the additional water vapour increasing the amount present in the air to the maximum possible at the prevailing temperature. Addition of water vapour sometimes acts at the same time as mixing, the two processes together causing cloud formation. It also acts at the same time as frontal and orographic lifting.

Additional water vapour is often supplied by evaporation of falling raindrops, the latent heat of vapourization being supplied by the air. Such cloud formed in rain will persist longest in stable air of high moisture content, sometimes for considerable periods after the rain has ceased. When there is a fairly strong surface wind, and the lowermost layers of air are well mixed, stratus or fractostratus is usually formed in rain as a quite uniform layer. With less wind, and consequently less mixing, a haphazard arrangement of clouds at widely varying levels may form in rain. In the absence of wind of sufficient strength to cause cloud, the air nearest the ground sometimes becomes saturated through the addition of water vapour, with consequent formation of fog. On account of the increase of water in the air, stratus formed in rain sometimes builds downward to form fog. The height of stratus formed in rain will depend upon the strength of the wind, the nature of the terrain, and the moisture content of the air. In the event of complete mixing, the cloud base is to be expected at the mixing condensation level. That mixing is not the only process operating in conjunction with addition of water vapour to cause formation of cloud in rain is indicated by observations made by the writer.

An interesting formation of cloud in rain was observed along the valley of the Gatineau River. The point of observation was on the west side of the river, and the cloud formation was observed against a background of trees on a hill on the east side. The wind being easterly, the lee side of the hill was visible from the point of observation. The precipitation was in the form of continuous light to moderate rain. Tenuous wisps of very thin cloud were observed to rise one after another from among the trees. These wisps could not be followed with certainty after they were above the trees, which grew all the way to the top of the hill. However, they appeared to go upward almost vertically as far as the tree tops, or a little higher. Although the motion beyond this point could not be observed with certainty, the wisps of fractostratus appeared to merge with the main body of stratus, which was very low and very ragged, extending down to approximately 50 feet above the top of the hill. The main body of this ragged stratus was estimated to be 400 feet above the point of observation, which point was well below the top of the hill. The rain was falling from a layer of nimbostratus, completely concealed by the stratus. The wisps of cloud were seen to stream out as they became high enough and far enough from the lee side of the hill to be affected by the wind. The question arises whether the cloud wisps were formed as fog among the trees and lifted by an upslope wind, or were formed in ascending damp air forced up by the wind. Whichever was the cause, this particular portion of the cloud was added from below, and was not formed at the top of a mixed layer. The amount of cloud observed in process of formation was actually very small compared with the total mass, and it is doubtful if this process is a really important one in the formation of low cloud in rain.

On the following day, in the same location, exactly similar cloud formation was observed. On this occasion light to moderate continuous drizzle was falling from an overcast of stratus based at about 500 feet above the point of observation. In the afternoon the drizzle became intermittent and cloud formation was observed to occur only during the time precipitation was actually falling. There was, on this occasion, a light northerly wind, very nearly down the river valley. The newly formed fractostratus was observed to rise slowly and drift with the wind as it rose. A band of fractostratus was seen against the hill, near the top, persisting for approximately ten minutes. This would argue that the anabatic wind was not very strong, which is quite likely in the circumstances, on account of the small amount of surface heating permitted by the thick cloud cover.

On a showery day, in the interior of British Columbia, the author observed, from the air, a very interesting cloud formation. Numerous vertical streamers of cloud were seen to extend from the ground to about 100 feet. Much fractostratus was observed in and near the showers at and below 1,000 feet.

In considering the formation of low cloud in rain it is interesting to speculate on whether or not the evaporation of the raindrops causes enough cooling of the air to assist materially in bringing about saturation. It is undoubtedly a contributing factor, but just how important a factor remains to be determined.

Cloud is also produced by lifting of fog of the type known as *steam fog*. This type of fog is common over water in seasons and at times of the day when the air is colder than the water, notably in the autumn and in the morning. Lifting of this steam fog by wind or by thermal updrafts or by both results in formation of low clouds. In the autumn and early winter this process commonly produces stratus cloud in the vicinity of the Great Lakes. The stratus so produced may remain over the water or may be blown inland. In some cases, it does not become cloud until after it has been blown inland.

The length of time stratus formed in this manner will persist after it has drifted away from its source will depend upon moisture and stability conditions in the air mass concerned. Over small lakes in the autumn after a clear night there is quite commonly an ascending column of steam fog spreading out into a mushroom form. Some or all of this steam fog becomes a small mass of stratus or fractostratus, which usually dissipates early in the day, because of surface heating and consequent updrafts. In most cases, the air contains a small amount of water vapour and such clouds evaporate readily without much heating. Rivers do not usually have a large enough water surface to contribute much to this fog and cloud formation process. The steam fog formed over open running water in very cold weather is usually of much too short duration to ever become cloud at all. In the case of a river, the valley, which serves as a collecting place for cold air, usually has more to do with fog formation and so indirectly with cloud formation than has the river itself.

FORMS TAKEN BY THE CLOUDS

The exact causes of the forms in which clouds are observed are at present imperfectly understood. These forms must obviously be the result of schemes or patterns of motion in the atmosphere. The clouds may show all of the motion in the air, or only a part. Whether they show all or only a portion is determined by the moisture content of the air, whether it is sufficient for condensation to occur throughout a deep enough layer to show the whole scheme of the air motion.

For some time the theory was generally accepted that Helmholtz gravity waves are responsible for the cloud forms. It was considered that the gravity waves were set up at the surface of discontinuity between layers of air of different densities. This theory is supported by the prevalence of cloud layers at inversion surfaces. It is also found to agree tolerably well with the forms of some of the undulated layer clouds, particularly stratocumulus. Some types of undulated stratocumulus have a well-marked wave form, with short crests and long shallow troughs on the lower surface and long crests and short sharp troughs on the upper surface. The waves are supposed to be formed perpendicular to the wind direction.

A theory of intersecting waves has been advanced as a means of accounting for the forms of some clouds. This, however, is rather difficult to explain. It would not seem to be possible unless one of the sets of waves originated at a distance in some place where the wind was different. That waves so formed would persist long enough to form a pattern by intersecting another set seems highly improbable.

Sir Gilbert T. Walker has stated that the forms of clouds can be quite adequately explained without any use whatever of Helmholtz gravity waves. By his theory, they are considered to be the result of a scheme of air motion in the form of cells. Each such cell has an updraft in the centre and downdrafts along the sides, and so has two cores rotating in opposite directions.

The forms displayed by some of the layer clouds, especially nocturnal altocumulus, lend strong support to the cell theory. The openings in the cloud may be very satisfactorily attributed to the downdrafts on the sides of the cells.

For the purpose of investigating cloud forms laboratory experiments have been made, using smoke as an indicator to show the air movements. The validity of applying the results of such experiments to the atmosphere appears to be open to doubt. The small scale of the apparatus employed and also the absence in the laboratory of many of the complicating factors present in the atmosphere are both quite powerful objections. Some of these complications take the form of condensation, evaporation, and sublimation of water, with the heat exchanges involved. Further, the nature of cloud components is rather different from that of smoke particles. There is no guarantee that water droplets and ice crystals in clouds will behave in the same manner in the atmosphere as does smoke in the laboratory apparatus. That such experiments do provide useful indications of the manner in which air may be expected to move under certain conditions is not questioned, but it appears dangerous to extrapolate the results too freely.

The form of *castellated* clouds indicates the presence of convective elements with cross-sections which are small in comparison with their heights. The towers usually form quite rapidly. Castellated clouds occur when the air has become conditionally unstable. It is thus one of the forms of greatest utility for an indirect aerological study of conditions aloft.

The *mammilated* forms of cloud have a characteristic appearance, and are highly significant. Explanations for their very striking form differ considerably. It is established that they occur only under conditions of considerable instability, and they are almost invariably associated with thunderstorm activity. Cumulonimbus mammatus has the mammilated surface on the underside of the anvil, and often it appears on the underside of the projecting portions of the sides of the cloud.

Humphreys suggests that mammilated cloud is formed by the projecting of snow portions of the upper part of a cumulonimbus above a layer cloud. The result of this would be downdrafts in spots, causing bulges in the cloud base, and hence the characteristic pouches of mammatus.

A. Wegener explained this form by assuming gravity waves. The sudden spreading out of an ascending cloud mass, such as cumulonimbus, under a strong inversion produces a situation where the air under this spread-out portion is dry and is colder than the air in the cloud. This makes for an appreciable density discontinuity, and formation of waves may be expected. In a cumulonimbus the strongly ascending cloud mass may force back a surface of inversion and so the rising air may pass its normal "equilibrium" position. The appearance of the pouches on the lower surface is then started by the returning movement of the "past equilibrium" air.

Another theory to account for the formation of mammilated cloud was advanced by J. Bjerknes. According to this theory, descending cold air slides along the under surface of the cloud concerned. The air is warmed by this descent sufficiently to become warmer than the air of the cloud immediately above. The colder air of the cloud then commences to flow downward, in the form of "pouches." Friction caused by these descending pouches stops the descent of the cold air in those parts in contact with the cloud surface. A species of "false equilibrium" is set up in this way.

ARTIFICIAL CLOUD

A form of cloud known as *artificial cloud* or *condensation trails* results from the passage of an aeroplane through the air under certain conditions. There are two effects produced by the passage of the aeroplane which correspond to the principal cloud-forming influences: cooling of the air and addition of water vapour. These are sometimes sufficient to cause condensation or sublimation, with resulting cloud formation. In order that this phenomenon may occur the air must be very nearly saturated. At the tips of the wings and propellors are vortices, in which there is a pressure reduction, causing cooling which may be sufficient to give rise to condensation or sublimation. The cloud so formed usually lasts for only a very short time, being therefore a very small cloud. The exhaust from the engine or engines contain a large amount of water vapour. This water vapour increases the moisture content of the air, sometimes to such an extent that condensation or sublimation occurs. The artificial cloud so formed persists rather longer than that formed by the wing tip and propeller tip vortices. Sometimes this type lasts as long as about 30 minutes. In both cases, the air contains little enough water vapour that the cloud formed by the aeroplane evaporates after a greater or less interval. When this occurs on the ground, as it does occasionally at very low temperatures when the relative humidity of the air at the surface is high, artificial fog forms.

4. Nuclei of Condensation and Sublimation

THE CHANGES of state of water in the atmosphere which result in formation of clouds are profoundly affected by the numbers and types of condensation and sublimation nuclei present.

When condensation to form clouds was first considered by scientists, no thought was given to the part played by nuclei. A notable advance was made in 1875 by Coulier and Mascart when they stated that nuclei were necessary for the condensation of water vapour in the atmosphere and hence were necessary for the production of fog and cloud. In 1880 a further contribution was made by Aitken. In discussing nuclei, Aitken used the term *dust,* but he recognized that the nuclei might be hygroscopic or non-hygroscopic in nature.

If air be admitted to a chamber and expanded, fog will form in the chamber as a result of the cooling caused by the expansion. The same air if progressively expanded produces successively decreasing amounts of fog with successive expansions. If the air is made to pass through a filter of packed cotton before entering the chamber, it is found that no fog forms. The decreasing amounts of fog with repeated expansions suggest that the air becomes "exhausted" of nuclei. The lack of condensation when filtered air is expanded indicates that the nuclei operative in this case may be removed by means of a mechanical filter. The filtered air becomes capable of producing fog and the "exhausted" air becomes again capable of producing fog if smoke is introduced into the chamber.

The theory of condensation without nuclei, described as *molecular condensation,* requires the joining of sufficient numbers of water molecules to form droplets. Against such condensation Simpson presents a very powerful argument. Since the diameter of a water molecule is 4×10^{-8} cm., about 5,000 molecules would be required to assemble together to make a drop with a diameter of 10^{-6} cm. It is quite unlikely that so large a number of molecules would come together by chance. In any case, a drop so formed would fly apart in any atmosphere having a supersaturation of less than 26 per cent. In the presence of nuclei, only sufficient supersaturation for condensation on those particular nuclei would be required.

Some physicists have advanced the theory that in the absence of other nuclei the ever-present ions in the atmosphere would serve. An unpublished theory of Langevin states that when condensation has been produced at enormous supersaturations directly from vapour and evaporation of the products of this condensation follows, there will remain droplets having radii of a few hundredths of a micron. Langevin's theory indicates that these remaining droplets are then capable of producing condensation at much lesser degrees of supersaturation. In this connection C. T. R. Wilson states that evaporation of a cloud produced by ionic condensation in purified air leaves charge-free nuclei. These charge-free nuclei can be only droplets of pure water. The suggestion has been advanced that these nuclei may be concerned in the formation of low clouds.

C. T. R. Wilson ionised by means of X-rays a sample of air which had been freed of ordinary condensation kerns, that is, freed of those nuclei which produce condensation in the neighbourhood of saturation. The sample of air so treated was then subjected to sudden expansion. Condensation did not appear until the expansion had proceeded to the point where the temperature had fallen 25 °C., and the relative humidity had increased to about 420 per cent. Condensation in this case was caused by the small negative ions acting as nuclei. For the small positive ions to become active as nuclei further expansion was required, sufficient to produce a cooling of 30°C., and a relative humidity of 600 per cent. Other investigations yield the result that condensation on positive ions requires a relative humidity of 790 per cent. The expansion involved is of the order of 1 to 1.38 or 1 to 1.4.

The results above mentioned were obtained by a laboratory experiment. In the atmosphere, small ions are always present, but in decidedly variable numbers. Near the earth's surface, the numbers range about 1,200 per cubic centimetre, the positive ions being slightly more numerous than the negative ions. The Wilson-Gerdien theory of ionic condensation was largely favoured for a time, but has since been abandoned. Kopp believed that a sufficient supersaturation for ionic condensation was to be found in the heads of cumulus clouds. This made the theory more attractive for a time. This idea is of considerable interest, in view of the great importance of the physical processes occurring in the tops of the clouds of vertical development. In such regions it is to be expected that a temperature drop of the order required for this ionic condensation would result in sublimation rather than condensation. This opens the question whether or not ions can serve as sublimation nuclei.

The rarity of occurrence of any appreciable degree of supersaturation in the atmosphere makes the enormous supersaturations required for ionic condensation decidedly unlikely. It is accordingly highly improbable that ionic condensation is an important factor in cloud formation.

NATURE OF THE NUCLEI

Some kind of condensation *kern* is indicated as virtually necessary for condensation without enormous supersaturation. Just what is the nature of these kerns is a most interesting problem, and is the subject of considerable controversy.

Non-hygroscopic particles do not appear to be particularly satisfactory as condensation kerns. In 1913 Wigand investigated an assortment of types of non-hygroscopic dust with a view to determining their fitness to serve as condensation nuclei. He measured the number of nuclei in a sample of air with an Aitken counter, and then added dust of various kinds: coal dust, floor dust, and dust from a beaten carpet. No appreciable increase was observed in the number of nuclei. This led to the conclusion that non-hygroscopic dust particles were not operative as condensation kerns. Boylan in 1926 carried out similar experiments, and confirmed the results obtained by Wigand.

The largest of the particles of non-hygroscopic dust are visible in the rays of sunlight. These large particles are found in the atmosphere in relatively small numbers. According to the results published by Wigand and Boylan, they have no part in bringing about condensation, even in supersaturated air. They can, however, be mechanically picked up by falling precipitation. On rare occasions sufficient numbers have been picked up to impart colour to the precipitation. The phenomenon of coloured rain has been known by various popular names, for example, "rain of blood." Reference is made in the *Iliad* to "bloody rain-drops." The smallest of such dust particles appear to be carried to considerable heights in the atmosphere, there to remain suspended for long periods of time. It seems a quite reasonable supposition that these particles may serve as nuclei for the formation of ice crystal clouds. In this connection, the question arises whether such non-hygroscopic particles are suitable sublimation nuclei but not suitable condensation nuclei. That such is the case appears highly probable. The fact that they are non-hygroscopic should not adversely affect their ability to act as sublimation nuclei.

An interesting possibility in connection with the action of non-hygroscopic nuclei is suggested by the vapour pressure differences over differently shaped liquid surfaces. The vapour pressure over a convex surface, such as that of a drop, is greater than that over a plane surface of the same liquid at the same temperature. For a concave surface, the vapour pressure is less than that over a plane surface. If the non-hygroscopic nucleus should take the form of a porous solid, water in its cavities would present a concave free surface, which would correspond to a lowering of the vapour tension. This would seem to suggest that non-hygroscopic but porous substances would make highly satisfactory nuclei.

Some revolutionary ideas about non-hygroscopic nuclei were published by Junge in 1936. He showed that he had successfully obtained condensation on particles of many kinds of non-hygroscopic dust. These included coal dust, earth, emery powder, ground-up clay, and floor dust. Condensation had also been obtained on small oil drops. The oil drops used were pulverized paraffin oil, and it was found that in order for them to act as nuclei it was necessary that they be very small. Junge came to the conclusion that the power of a particle to act as a condensation nucleus does not depend upon the material of which the particle is composed. He does not specify whether all substances are equally satisfactory as nuclei, or whether some require more supersaturation than others.

Hygroscopic substances exhibit a chemical attraction for water, which encourages condensation of water vapour on them in preference to other substances. This same attraction is capable of causing condensation in an atmosphere which is not saturated. It appears that hygroscopic substances possess in general more satisfactory qualities as nuclei than do non-hygroscopic substances.

With a view to gaining information regarding nuclei, investigators have evaporated cloud droplets. They have so far failed to find any observable residue, even with the most powerful microscopes.

The exact form taken by hygroscopic nuclei is most frequently that of droplets of solution. It was established by Lord Kelvin that the vapour pressure over the surface of a water drop in equilibrium with its surroundings is greater than that over a plane water surface. This vapour pressure difference is inversely proportional to the radius of the drop. The vapour pressure over a drop is then greater than the *saturation* vapour pressure for the liquid concerned, since the saturation vapour pressure is defined as that over a plane surface. A vapour pressure greater than the saturation value would then be required in the vicinity of such a drop if its evaporation is to be prevented. For *condensation* to occur on the drop, a higher degree of supersaturation must prevail. The vapour pressure over the surface of a solution is always less than that over pure water at the same temperature. For this reason, a solution is hygroscopic, and absorbs water vapour from the surrounding air, becoming diluted in the process. From this it follows that nuclei in the form of droplets of solution may be expected to be satisfactory condensation kerns. When particles of hygroscopic substances are exposed to sufficiently damp air, they become droplets of solution.

It has been suggested that the nuclei of condensation are primarily of a gaseous nature. This theory supposes that they are composed of oxides which combine with the water vapour of the atmosphere to form nuclei in the form of acid drops. Such acid drops would be hygroscopic. The presence of these oxides in the atmosphere is attributed to the action of sunlight on the gases of the atmosphere, and to pollution of the atmosphere by the burning of coal and by other industrial processes.

It is highly probable that there are separate and distinct *sublimation nuclei*. Bergeron says, "Theoretically, that is, under ideal, undisturbed conditions crystallisation will even demand more special properties of the nuclei than condensation (solid particles showing the angles of the hexagonal crystal system), i.e. will not take place in lack of such nuclei. In the air, there will, however, probably be a small amount of such particles, which can gradually get into action as sublimation nuclei, as the temperature falls."

That the sublimation process will require nuclei having properties different from those of condensation nuclei appears entirely reasonable. Some kind of insoluble particles appears to be indicated as the most likely sublimation nuclei. If the nuclei present were drops of solution they would certainly not induce sublimation. Consider for example a nucleus consisting of a crystal of salt from the sea on which water has condensed. Such a crystal surrounds itself with a saturated saline solution which does not readily freeze. If freezing does not take place before the nucleus has completely dissolved it will be almost impossible afterwards, for it will remain supercooled in the absence of a solid particle to promote freezing. At very low temperatures, below −20°C., the freezing precedes the dissolution of the salt. Normal snow forms at low temperatures and the formation of prisms and needles is attributed by Stüve to insoluble nuclei, their formation on such nuclei being possible at any temperature below 0°C.

Minerals such as quartz, having the same hexagonal crystal structure as ice, might make satisfactory sublimation nuclei. Ice crystals already present in the air would, of course, make highly satisfactory nuclei for additional sublimation.

The thin and diffuse character of the ice crystal clouds may perhaps be ascribed to a scarcity of sublimation nuclei at the levels at which these clouds are formed. However, it seems a much more acceptable theory that the diffuse character of the clouds is caused by the smallness of the supply of water vapour available in air at the low temperatures which prevail where ice crystal clouds form.

The size of the nuclei appears to be an important consideration. The vapour pressure over a liquid drop is greater than that over a plane surface of the same liquid at the same temperature, and increases as the size of the drop diminishes. Greater supersaturation is therefore required for condensation on small drops than on large ones. For example, for condensation on a drop of pure water having a diameter of 10^{-2} cm., a supersaturation of .002 per cent would be required, while for condensation on a drop having a diameter of 10^{-6} cm., supersaturation of 26 per cent would be required. It thus becomes evident that condensation will occur more readily on the larger droplet nuclei than on the smaller ones.

There appears to be in some cases a factor favouring the smaller nuclei. It has been suggested that while the relatively large nuclei must be hygroscopic, the very small ones are not necessarily hygroscopic. In 1936, Junge obtained condensation on very finely divided paraffin oil, which would seem to bear out the contention regarding non-hygroscopic nuclei of very small size.

The size of effective sublimation nuclei must also be considered. These will quite surely be solids, and there appears to be a lower limit to the size of such solid particles. According to Green the lower limit of the size of such "disintegration aerosols" is of the order of 10^{-5} cm. The size of the average condensation nucleus was computed by Köhler to be of the order of 0.1 micron, with a mass of 10^{-15} grams. The following table, after Simpson, shows the various degrees of supersaturation required for condensation on nuclei of different sizes, and of hygroscopic and non-hygroscopic characters.

Diameter of nuclei requiring various supersaturation for condensation

Supersaturation (per cent)	0.5	1	2	3	5	10	20	
Non-hygroscopic		46	23	12	8	5	2.4	1.3×10^{-6} cm.
Hygroscopic		16	8	5	3.4	2.2	1.3	$.8 \times 10^{-6}$ cm.

(d at 100 per cent R.H.)

The electric charge carried by a droplet nucleus has been suggested as one of the attracting forces causing condensation. However, it has been shown experimentally that the charges on nuclei seldom exceed one electronic charge. A charge of many times this magnitude would have only an infinitesimal effect on the hygroscopic properties of the nucleus. It seems very doubtful if this effect is of any great importance.

SOURCES OF THE NUCLEI

While it is quite probable that nuclei of condensation and sublimation come from a number of sources, there is little doubt that the vast majority will come from a very few of these sources.

The solid particles of non-hygroscopic dust found in the atmosphere come from several sources, some of which are strictly natural and some artificial. The mineral dust resulting from disintegration of rocks provides a source of small particles. This dust is picked up by the wind, and it is reasonable to assume that it is sometimes carried to quite high levels. If the particles are small enough, they will settle slowly and so will remain in the air for a fairly long period. Pollen grains from plants are carried about by the wind, but the pollen supply is seasonal.

Volcanic eruptions have been held responsible for the presence of dust particles at unusually high altitudes. Crepuscular rays indicating the presence of dust at high levels were observed in Europe in 1883 and 1884, and then again in 1903 and 1904. These times show an interesting relation to two major volcanic eruptions: that of Krakatoa on August 29, 1883, and that of Mount Pelee on April 25, 1902. These particles consist of cinders and other dust. It has been assumed that the dust projected into the atmosphere in this manner takes a period of years to settle to earth.

Among the solid non-hygroscopic dust particles in the atmosphere must also be numbered the solid particles in smoke resulting from the imperfect combustion of fuels. Such solids are found in the smoke from bush fires and forest fires, which may be considered a more or less "natural" source. Solid particles are also found in smoke produced by the burning of fuels in furnaces. This must be considered a strictly non-natural or "artificial" source. Coal smoke contains many particles of carbon which are often stuck together with bits of tar, so forming fairly large irregular masses. These irregular masses occur in relatively small numbers.

In 1935, Junge produced non-hygroscopic nuclei, probably composed of insoluble oxides of zirconium and yttrium, by drawing air over the glowing filament of a Nernst lamp.

There is an impressive array of evidence in favour of the theory that the vast majority of the nuclei of condensation are composed of salt. Rime and cloud droplets were studied by Köhler at mountain observatories and the results of his analyses indicate that the chloride content is quite constant. The chloride content in rain appears to be consistent in various parts of the world, and to agree with Köhler's determinations directly from cloud droplets. As a result of these studies Köhler arrived at the conclusion that cloud particles form on nuclei containing 1.847×10^{-14} grams of sea-salt. According to Wright, in the formation of fogs deposition of water does not occur on combustion nuclei if sea-salt nuclei are present. Wright also states that sea-salt nuclei are larger than combustion nuclei. This larger size results in a lower vapour pressure over their surfaces, and makes them more satisfactory condensation nuclei. The size of a sea-salt nucleus is a function of the relative humidity and the mass of sea-salt in the nucleus.

When waves of the sea break on the shore, or wavelets overturn or foam bubbles break, aerosols are produced by the mechanical dispersion of the water. The droplets so produced vary considerably in size, from the smallest possible droplet to large droplets which fall back into the sea immediately. It is probable that there is a favoured size. Findeisen objects that the particles so produced cannot be smaller than those produced in the relatively violent blast of air employed in a spray bottle. He states further that "no known method" of spraying can produce droplets smaller than 10^{-5} cm. in diameter, and that the majority produced by ordinary methods are much larger. According to Wright these relatively large droplets could not remain in suspension long enough to serve as nuclei. The spray bottle analogy does not take into account the effects of bursting bubbles and of impact on the shore. Simpson objects that the number of sea-salt nuclei which can be produced in this way is insufficient to account for the normal rainfall over the earth.

While the largest drops produced by sea spray fall quickly back to the surface of the sea, the smaller ones are carried for considerable distances. These small droplets are capable of remaining in suspension for periods of time sufficiently long to ensure a wide distribution. The limiting fall velocity in calm air of a droplet of water of radius r between 0.1 micron and 100 microns is approximately $10^6 r^2$ cm. per second, (for r in cm.). A cloud droplet having a radius of 10 microns falls at the rate of only 1 cm. per second, and very weak updrafts are sufficient to hinder its descent or even to carry it up again. A droplet nucleus having a radius of 0.1 micron descends somewhat less than 10 cm. per day.

While the slow rate of fall of the smallest sea-salt droplets makes possible a wide horizontal distribution, there is considerable doubt about the extent of the vertical distribution. In 1938 the presence of sodium atoms at altitudes of the order of 60 km. was demonstrated by the identification of the D line in the spectrum of the night sky. It has been suggested that these sodium atoms might be traced to ocean particles which had attained heights considerably greater than had previously been suspected. This would make possible the functioning of salt nuclei in connection with formation of mother-of-pearl clouds.

While the droplets of sea spray are being carried about in the air there must be a certain amount of evaporation from their surfaces. This evaporation will depend upon the dryness of the air. If the relative humidity of the surrounding air is less than 98 per cent, at least part of the droplets may be expected to evaporate. The evaporation may be expected to continue until the concentration of the solution is such that a balance is reached. If the relative humidity of the surrounding air is sufficiently low, the droplet may be reduced finally to a particle of dry salt. For sodium chloride this condition would be brought about at a relative humidity of 74 per cent. This stage would be reached at slightly different relative humidities for the other salts present in sea water. It has been suggested that there may be some subdivision of the salt particles after desiccation, and that the resulting very small crystals may be carried to high altitudes. According to Owens, the diameter of a drop of sea spray cannot be reduced by evaporation to less than one-quarter of its original value. This would place a lower limit on the size of nuclei obtained from sea spray.

Findeisen objects that the sea spray droplets are not small enough to have a salt content of the right order of magnitude. He states further that sea-salt is present in the atmosphere in the form of relatively large particles which either form the nuclei for a few of the cloud drops or become absorbed on drops or ice crystals already formed around "other nuclei." The presence in the atmosphere of fairly large particles of sodium chloride is demonstrated by the yellow flashes commonly observed in the flame of a Bunsen burner. However, the number of these particles seems to be much too small to account for the condensation which takes place in the atmosphere, or for the salt content of precipitation.

The suggestion has been advanced that salt nuclei might come from the surface of the salt-encrusted deserts. This solution, however, does not appear satisfactory. The salt-covered deserts cover only a relatively small area, and the oceans cover five-sevenths of the surface of the earth.

An attractive solution for the problem of the origin of salt nuclei was offered by Melander in 1897. He found that small quantities of salt were deposited on microscope slides exposed over warm salt solutions. The hypothesis was formulated that the water molecules leaving the surface of a salt solution during the process of evaporation at ordinary temperatures carry with them minute quantities of solution. This might be the explanation of the origin of nuclei, since it does not depend upon strong winds causing spray. The process could be expected to go on continuously and so would provide an ideal source of nuclei. Some coalescence would be required following this initial production of droplets if droplets of the size now recognized as nuclei are to be produced.

Gases enter the atmosphere in smokes of various kinds. In coal smoke there are present considerable quantities of sulphur dioxide, which is often oxidized under the influence of sunlight, becoming sulphur trioxide. The highly hygroscopic sulphur trioxide combines with water vapour present in the air to form minute droplets of sulphuric acid, which will be very efficient hygroscopic nuclei. Other industrial processes will produce gases which may cause formation of acid nuclei. In this connection, it is interesting to note that Aitken recognized that the activity of ordinarily active condensation kerns was greatly increased by the action of sunlight.

It is held by some investigators that hydrogen peroxide and ozone are produced by the ultra violet rays of the sun from the water vapour of the atmosphere. These then act as oxidizers for the process outlined above.

It is suggested by Simpson that nitrous acid is formed naturally in the atmosphere from nitrogen, oxygen, and water vapour by lightning and by ionisation caused by cosmic rays and radioactive substances. It is also possible that ozone diffuses downward from the ozone layer in the upper atmosphere. Coste and Wright consider nitrous acid a much more likely constituent of the atmosphere than sulphuric acid. In 1935 they examined the production of nuclei by flames, which they believed to be a source of large numbers of nuclei. The combination in the flame of nitrogen and oxygen of the atmosphere with some of the water vapour is believed to produce nitrous acid.

In 1935, Junge produced hygroscopic nuclei, which he pronounced to be sulphuric acid, by burning a small jet of coal gas in a current of air. They were produced from the sulphur in the coal gas.

Mineral dust appears a promising source of sublimation nuclei. Dust produced by the distintegration of rocks and blown to fairly high levels is a probable source of supply. Sand from desert regions is another possibility. It is likely that both the mineral dust and the sand contain suitable minerals, such as quartz, which have a hexagonal crystal structure. In this connection, it appears rather doubtful if there would be a sufficiently wide distribution to account for ice crystal clouds in all parts of the world. Areas which have recently acquired the characteristics of deserts must be left out of consideration, as ice crystal clouds have existed in considerable quantity for a very long time.

That sublimation nuclei have a cosmic origin is a rather attractive theory. Of the possible cosmic sources, meteoric dust appears the most likely. If this is the case, meteors and the dust from them must have a fairly uniform distribution. The dust would also have to settle in a uniform manner. Such very fine dust particles as those which would be involved may be expected to settle slowly, thus remaining available as sublimation nuclei for a considerable length of time.

The possibility of spontaneous solidification into ice of some of the super-cooled droplets in a cloud suggests another source of sublimation nuclei, namely the ice crystals so formed. This would require disturbed droplets, such as those in a turbulent cloud. Those droplets which are larger than a certain size crystallize more readily. This certain size is called by Köhler *Gefriergrösse*. This process could, of course, operate only in heap clouds where both water and ice phases exist, preceded by a water phase only.

Certain conclusions regarding the nature and sources of the nuclei of condensation and sublimation appear justified. The exact nature of the nuclei and of the processes which produce them are problems as yet not completely solved. The question remains whether they are salts or acids. If they are acids, whence comes the chloride content of rain and of cloud droplets? It could, of course, be the result of salt particles in the atmosphere being picked up by cloud droplets and raindrops. This would require a wide and uniform distribution of salt in the atmosphere. If there is such a wide and uniform distribution of salt, why should the salt particles not be the nuclei? The necessity for postulating chemical processes in the atmosphere would then be removed.

The nuclei need not necessarily all be of the same material. Certain properties, however, appear at least highly desirable if not entirely necessary. Nuclei of a hygroscopic nature are definitely more effective in promoting condensation than are those of a non-hygroscopic nature. Sublimation nuclei of hexagonal crystal structure are more effective than others. Sublimation nuclei must also be solids.

It is necessary that the nuclei be produced by entirely natural sources, or at least that a very large proportion of them be so produced. To ascribe to industrial processes, such as the burning of coal, the production of nuclei appears most unwise. Such processes are sometimes responsible for the formation of local fogs, but for large scale cloud production they are not adequate. There is, further, adequate geological evidence that cloud and rain existed long before coal. Else, whence came the rain which made possible the growth of the vegetation which made the coal? All forms of artificial pollution of the air must be left out of account in seeking the sources of condensation and sublimation nuclei. The natural processes responsible for production of condensation and sublimation nuclei must have been in operation over a very long period of time.

Further, the processes which produce nuclei must operate continuously. They must not depend upon special sets of circumstances. Non-continuous processes may reasonably be considered responsible for the production of nuclei for rare types of cloud, such as those occurring in the stratosphere. Such non-continuous processes as volcanic eruptions, bush fires, etc., must be considered to be supplementary sources of nuclei only, if indeed they are sources of nuclei at all.

From the point of view of the continuity of the process, Melander's hypothesis provides a much more attractive solution to the problem of the origin of salt nuclei than does the theory that they result from sea spray.

Continuous processes must also be considered responsible for production of the nuclei of sublimation. Deserts and high winds do not fulfil this requirement. While this combination certainly produces the rare phenomenon of coloured rain, it should not be held responsible for the production of ice crystal clouds. Some process resulting in a wider and more uniform distribution of sublimation nuclei is necessary. Dust of cosmic origin appears to be a much more probable source. This could be produced by continuous processes, and it is reasonable that the particles would be small enough to settle very slowly, becoming well distributed during the settling.

5. Natural Precipitation

A DISCUSSION of the relation between clouds and precipitation is best begun with a clear idea of the difference between a cloud droplet and a raindrop. Both cloud droplets and raindrops fall, relative to the air, but the rate of fall of cloud droplets is so small that they give the appearance of floating. While the droplets in the clouds fall relative to the air, they do not necessarily fall relative to the earth, on the contrary, updrafts may cause them to rise.

Some exact criterion for distinguishing cloud droplets from raindrops would be useful. To say that raindrops fall and cloud droplets float is not satisfactory, nor would it be satisfactory even if it were literally true. The rate of fall of drops depends upon their size, and their rate of fall relative to the earth depends also upon the strength of the updrafts in the air. Various arbitrary limits have been suggested for the size of a raindrop. Hann adopts 0.12 mm. as the minimum size of a raindrop, without giving any reason for the choice. Another value suggested was 0.07 mm., because the rate of fall of drops of this size is one metre per second and this is sufficient for the drop to fall through the usual ascending currents of rain clouds. Findeisen pointed out that the physical difference between a cloud droplet and a raindrop is that a raindrop must be of sufficient size to reach the ground without completely evaporating. This fixes the limiting size quite exactly. Findeisen has derived a formula showing that the distance travelled in an unsaturated atmosphere varies as the fourth power of the radius of the drop, and produces this table:

(Pressure 900 mb., temperature 5°C., relative humidity 90 per cent)

Radius of drops (cm.)	*Distance of fall before evaporation*	
10^{-4}	3.3×10^{-4} cm.	cloud droplets
10^{-3}	3.3 cm.	
10^{-2}	150 metres	
10^{-1}	42 km.	raindrops
2.5×10^{-1}	280 km.	

From the table it will be seen that a drop needs to have a radius of 2×10^{-2} cm. or more if it is to reach the ground from a cloud at any of the usual cloud heights. This is the value adopted by Findeisen as the lower limit of the size of a raindrop. It is evident that slightly smaller drops will reach the ground from extremely low clouds.

FORMATION OF PRECIPITATION

For many years there was no real attempt on the part of meteorologists to determine what specific processes in the clouds produce precipitation. Condensation and precipitation were regarded as merely successive stages in the same process. However, it is now recognized that such is most certainly not the case. What processes cause the release of precipitation from a cloud? Obviously, something must occur which disturbs the stability of the cloud, and so causes growth of drops and release of precipitation. As demonstrated by Wigand and Schmauss, a cloud may be regarded as a colloidal suspension of water in air, an aerosol, similar in many respects to a hydrosol. Precipitation processes must be capable of breaking down the colloidal stability of the cloud. Bergeron discussed the following five factors which tend to maintain colloidal stability, and also the results of their failure.

1. Uniform, that is, equal and unipolar *electric charge* on the cloud droplets. The repulsion operating between like charges would then hinder coalescence. Wigand and Frankenberger demonstrated that fogs with high electrical charges of the same sign on the droplets remained "dry," while fogs without charge were "wet" or drizzling fogs. The high and uniform charge in this case is enough to prevent coagulation of the fog droplets into droplets of drizzle size. However, it seems certain that unless some other process has already produced coalescence, the available strength of attraction due to electrical charges of opposite sign is insufficient to make an appreciable contribution to the growth of drops. In other words, electrical attraction in itself appears to be incapable of producing the initial release of precipitation.

2. Uniform *size* of cloud elements. The capillary and hygroscopic forces which would be put into operation by differences in size of adjacent cloud droplets do not appear to be adequate to start precipitation. During the formation of cloud droplets, this effect operates to an appreciable extent. Once the droplet has grown to a radius of more than 10^{-4} cm. the effect becomes insignificant; and as Bergeron puts it, "for the formation of ordinary raindrops or snowflakes this effect works some billion times too slowly." There is still a tendency for the larger droplets to grow at the expense of the smaller, but it will not be a major factor in releasing precipitation. Once precipitation has been started by some other means, the effect of differences in sizes of droplets will join with other effects in augmenting the precipitation.

3. Uniform *temperature* of cloud elements. If adjacent cloud droplets could have a sufficiently large temperature difference, the colder drops would grow at the expense of the warmer. The vapour tension over the warmer drops would be higher and would tend to make the vapour tension in the surrounding air greater than that over the colder drops. This would lead to a transfer of water from the warmer drops to the colder ones through the vapour. This effect, how-

ever, does not appear to be adequate to cause the initial release of precipitation. A sufficient temperature difference between the cloud elements would exist only if they came from different parts of the cloud, and were brought together as a result of turbulence or of a difference in their rates of fall. Turbulence is, in general, likely to diminish temperature differences. Differences in fall velocities would occur only if the drops were of different sizes. In order that differences in size may exist, some of the drops must be caused to grow by some other process. Different temperatures in different portions of the cloud are sometimes brought about by radiational cooling of the upper portions. Reynolds considered that radiational cooling of the upper surface of a cloud was the means by which precipitation was begun. According to Reynolds' theory, the droplets at the top of the cloud are cooled by radiation and fall through the cloud, picking up additional water by coalescence with other droplets as they fall. In daylight, however, the effect of radiation is to warm the upper surface of the cloud, but cooling by radiation does occur after sunset. This theory, then, appears to explain satisfactorily the very light precipitation and virga from stratocumulus in the evening and at night.

4. Uniform *motion* of cloud elements, preventing fusion by collision. Turbulence is capable of causing collision between droplets with the possibility of coalescence. However, surface tension strongly opposes this fusion, the droplets having a tendency to rebound from each other rather than to coalesce. It does not appear that coalescence by collision could possibly be the principal means of releasing precipitation.

5. Uniform *phase* of cloud elements, that is, all elements liquid water or all elements ice. This is the effect which Bergeron considers all-important.

The large difference between the vapour pressure over ice and that over water at the same temperature results in the vapour tension in air in which both water droplets and ice crystals are present being higher than that over the ice and lower than that over the water. Accordingly, sublimation will occur on the ice and evaporation will take place from the water droplets. The result is a rapid transfer of water from the droplets to the crystals, leading to considerable growth of the crystals.

In these circumstances growth of ice crystals is very rapid, and thus the presence of ice crystals in a cloud formerly free from them will start a rapid release of precipitation. Of this ice and water effect Bergeron says, "almost every real raindrop and all snowflakes originated around an ice crystal."

The ice crystal effect is well illustrated by the change from cumulus to cumulonimbus. In temperate latitudes it is a matter of common experience that this change is clearly shown by the glaciation of the top. As soon as its top becomes "infected" with ice crystals, the cloud becomes much more active. The importance of the ice crystal effect is very well illustrated by the fact that when the freezing level is low much less vertical development of the cloud is required for the production of precipitation. Another illustration is supplied by the warm front cloud systems, in the altostratus-nimbostratus combinations.

Once the release of precipitation has been started by the ice crystal effect, other conditions making for colloidal stability are disturbed and additional effects contribute to the increase of precipitation. The effects of size difference and of relative motion of droplets become appreciable and cause further growth of the droplets. It will be found that the factors already discussed all come into play once the precipitation is released.

The ice crystal theory, to be entirely satisfactory, must explain precipitation in all parts of the world and under all conditions. Observations provide numerous exceptions. In middle latitudes, the writer has observed precipitation in the form of real raindrops falling from clouds which did not appear to reach freezing level. Rain has, in fact, frequently been observed to fall from quite thin stratocumulus, topped far below the freezing level.

Simpson points out an interesting feature of the argument for and against the ice crystal theory. In Europe, and in temperate latitudes generally, most of the clouds which produce precipitation in appreciable quantities are thick enough to reach above the freezing level, and the fact that they do penetrate this colder region does not prove such penetration to be necessary for the production of precipitation. In warmer latitudes, clouds must be much thicker before they extend above the freezing level, and it has been amply demonstrated that they do not need sub-freezing temperatures to produce precipitation.

There must be some precipitation release which does not depend upon the simultaneous presence in the cloud of the two phases, water and ice, and this process must be one which operates at higher temperatures. Direct condensation on droplets does not appear to be the answer, since the time required for them to grow to raindrop size would be excessive. Nor does coagulation by collision appear to be the initial release process, although once there are differences in size from other causes, the difference in rates of fall will greatly increase the probability of collision.

A very satisfactory explanation for the release of precipitation in the absence of ice crystals has been supplied by Petterssen. At higher temperatures, very small temperature differences will give rise to considerable vapour pressure differences, and will cause growth of droplets and release of precipitation in much the same way as in the presence of ice crystals. At higher temperatures, then, there is no need of a phase difference between neighbouring cloud components to serve as a precipitation release.

It is further pointed out by Petterssen that if there is considerable turbulence, droplets of different temperatures will come together and the precipitation process will be started thereby. Once the release has begun, electrical charges will assist to some extent. In order that the droplets may become large, there must be a considerable thickness of cloud, the droplets growing as they fall.

While the ice crystal theory adequately explains the release of precipitation in a large number of cases, it does not appear to be the only process releasing precipitation. Temperature differences between adjacent droplets, brought together by turbulence, seem to account for showers at higher temperatures. However, in some of the cases where real raindrops have been observed to fall from clouds whose tops are far below freezing level, the cloud is too thin for much temperature difference between top and bottom and there is very little turbulence. The temperature difference required at higher temperatures being very small, this objection is actually reduced to an objection on the ground of insufficient turbulence. Precipitation occurring in the evening from strato-cumulus can, in some cases, be very satisfactorily explained by the operation of the Reynolds effect.

No proof has so far been advanced to establish any one process as a universal release of precipitation, and it appears unlikely that there is any. It is the writer's belief that different effects operate under different conditions.

PRECIPITATING CLOUDS

Certain of the cloud forms are almost always precipitation clouds; other forms are never precipitation clouds; and still others are occasionally precipitation clouds. The very diffuse forms of cloud which are composed exclusively of ice crystals are not precipitation clouds. These include cirrus and cirrostratus, although some of the cirrus forms have "tails" on them, which tails may be considered to be a form of virga. Cirrocumulus is not a precipitation cloud.

Altostratus, in its thinner forms, is not a precipitation cloud. Rain or snow in appreciable quantities, and of rather a wide range of intensities, falls from the thicker forms of altostratus. From the higher and thinner forms of altostratus precipitans, precipitation is mostly in the form of virga, small amounts reaching the ground in the form of intermittent rain or snow. Altocumulus rarely produces any precipitation other than virga. Layers of middle cloud composed of a mixture of altocumulus and altostratus commonly act as precipitating clouds.

The thickening which occurs when altostratus becomes nimbostratus gives rise to an increase in the intensity of the precipitation. There are very few precipitation-free periods when nimbostratus is present. In nimbostratus there is sufficient turbulence to sustain raindrops until they grow to fairly large sizes, and also to cause a considerable number of collisions.

In many cases, no precipitation falls from stratus, but sometimes it produces drizzle or very light snow. It is not thick enough nor turbulent enough to produce precipitation in the form of drops larger than drizzle. Stratocumulus is not generally a precipitation cloud, but occasionally very light rain or snow falls from it. It is not so smooth as stratus, but the turbulence in it is only slight. Sufficient updrafts, however, are present to support drops of a larger size than drizzle drops. Fractostratus and fractocumulus are too thin to produce precipitation. Rain falling from altostratus and nimbostratus, and forming fractostratus as it falls, may sometimes appear to come from the fractostratus, whereas it actually falls through the fractostratus.

When rain or snow falls from clouds of vertical development, it is in the form of showers. No precipitation whatever falls from cumulus humilis. Cumulus congestus sometimes produces showers. Heavy showers fall from cumulonimbus, and the occurrence of such showers is adequate evidence of the presence of cumulonimbus, even though the distinguishing features of cumulonimbus be concealed by other clouds. It is the only cloud which can produce real hail.

Formation of hail is caused by falling drops of rain being arrested by the updrafts in the cloud and carried to higher and colder levels. At these higher and colder levels the drops become frozen, and while still wet they are carried high enough in the cloud to pick up ice crystals. The growing hailstone falls off the edge of the updraft which has been supporting it, and goes down through the lower portions of the cloud, picking up water as it falls. This water freezes, and forms another layer of clear ice. The process is repeated until the hailstone becomes heavy enough to fall through the updrafts to the ground. It may melt before it reaches the ground, and arrive as a raindrop. The sizes of the hailstones falling from a cumulonimbus cloud are determined by the strength of the updrafts in that particular cloud. Hail may fall from the under side of the anvil of a cumulonimbus, as well as from the base of the cloud.

Occasionally, light snow falls without previous cloud formation, the snow falling as it forms. Such snow is usually in the form of quite small ice crystals, and there is some basis for the contention that the term sometimes used to describe it, namely "ice crystals," is more correct than "snow." This usually occurs in very cold weather. Very occasionally, flakes of snow are formed in this manner, the crystals becoming interlocked. This occurs when the temperature is not quite so low.

6. Induced Precipitation

AN INVARIABLY successful process for producing rain and snow would have tremendous significance for virtually everyone. Hence, recent experiments in the production of man-made precipitation have aroused wide interest.

The basic requirement of a rainmaking process is that it should disturb the colloidal stability of a cloud before the latter would be disturbed by natural means; or that it should disturb the colloidal stability of a cloud which would otherwise remain stable. A really complete rainmaking process would also produce the cloud itself.

Methods for disturbing the colloidal stability of clouds have engaged the attention of several investigators in recent years. There have been speculations regarding the possibilities of severe concussions being effective; but this forms no part of the current experiments. The crucial point in the Bergeron theory of precipitation release is the simultaneous presence, in a supercooled cloud, of ice and liquid water. So, if ice crystals could be introduced into a supercooled cloud not naturally containing ice crystals at the time, rain might possibly be caused to fall; or, if the temperatures were too low for rain, snow might fall. Some method for producing these ice crystals in the cloud then seems indicated. If some of the supercooled droplets could be induced to freeze, or if ice crystals could be caused to form by sublimation, the required water and ice mixture would be attained. The introduction of some very cold substance into the cloud has promising possibilities. This means of producing rain was tried in Holland in 1930. Recently, many experiments have been performed based on this principle. The very cold substance used is solid carbon dioxide, known commonly as "dry ice." The temperature of this substance is $-80°C$. When solid carbon dioxide, in pieces of suitable size, is introduced into a cloud of supercooled water droplets, numerous ice crystals of very small size are formed by sublimation of some of the water vapour in the cloud air. The presence of these ice crystals may start a transfer of water from the droplets to the crystals, through the vapour, in the same manner as in the release of naturally formed precipitation. Other inoculants have also been used. Silver iodide, which has a crystal structure similar to that of ice, is used to provide nuclei to encourage sublimation of some of the water vapour in the cloud air. It has been thought by some experimenters that the silver iodide crystals may remain in the air, if not immediately used as nuclei, and become effective at a later time; whereas the solid carbon dioxide evaporates to the gaseous form rather quickly. Water drops have also been used, with a view to having them grow by picking up cloud droplets as they fall, and possibly initiating a chain reaction.

The "chain reaction" postulated for the inoculation of a cloud with water drops would be caused by the falling drops with which the cloud had been inoculated, growing to such size that they would be broken up by the air stream. Each of the resulting drops is assumed to sweep up the cloud droplets in its path, and to grow in turn to such a size that it breaks up, and so on as long as is permitted by the cloud depth. Recent experimental and theoretical work by Swinbank indicates that coalescence of cloud droplets is unlikely, and that coalescence of cloud droplets with drops falling through the cloud is also unlikely.

The method most often used for inoculating clouds is to release the inoculant from an aircraft flying above or in the cloud being treated. When solid carbon dioxide is used, it is broken into small pieces, preferably about three-eighths of an inch in diameter. Sometimes it is broken with hammers, and sometimes with crushing machines. The rate of "seeding" is a question which has been and still is open to discussion. Dr. V. J. Schaefer maintains that many investigators overseed the clouds, and so fail to obtain the best results. It is his belief that overseeding results in too many ice crystals, none of which can grow large enough to fall. However, some of the other experiments do not seem to indicate that there is any critical seeding rate. The experiments conducted by the National Research Council of Canada do not show any critical value, nor did the few in which the author participated with the Hydro-Electric Power Commission of Ontario.

Experiments have been performed in many widely separated locations, by many individuals and organizations. The pioneers in the present extensive experimentation are Langmuir and Schaefer of the General Electric Company, and their associates. An extensive and admirably instrumented and performed series of experiments has been carried out by the United States Weather Bureau under the direction of Gunn, Coons, and Gentry. Some very interesting and significant experiments have been performed in Hawaii by Leopold and Halstead. The Australian Council for Scientific and Industrial Research has conducted a number of experiments, and has reported some spectacular results. In Canada, the National Research Council has performed a number of experiments at Arnprior, Ontario, and Suffield, Alberta. These experiments have been performed with the co-operation of the Royal Canadian Air Force and the Meteorological Division, Department of Transport. Attempts were made in the spring of 1948 to extinguish forest fires in Northern Ontario by rainmaking. The Ontario Hydro-Electric Power Commission, with the assistance of the author, carried on a small series of experiments in the fall of 1948 at Kapuskasing, Ontario.

Experiments so far performed indicate that, under certain conditions, the production of precipitation and the modification of clouds is definitely possible. The conditions for success are much the same as those for production of natural precipitation. In some cases, the precipitation produced occurred without any natural precipitation in the close vicinity; and in other cases there was natural precipitation near by. As is to be expected, when the seedings were confined to the type of cloud which is most likely to rain, the proportion of successes was higher· then when all types of cloud were seeded. The amount of rain produced varied, and was, in most cases, rather small. In experiments carried out in colder weather (including those in which the author participated) the precipitation was in the form of snow. On some occasions, rain has been caused to fall at a heavy rate. In many cases, no means were available for measuring the precipitation, and whether it consisted of rain or snow at the ground, or only of virga, had also to be determined as well as possible by observations from the aircraft.

In the Canadian experiments, inoculation of clouds at above freezing temperatures resulted in subsidence of the cloud only. This might well have been an aerodynamic effect produced by the passage of the aircraft, and not related to the inoculation of the cloud. The precipitation obtained from stratus clouds was very light and often did not reach the ground. That from cumuliform types of cloud reached the ground more often. However, more cases of precipitation without natural precipitation near by occurred from the stratus clouds than occurred from the cumuliform types. This is to be expected, since the stratiform types, with the exception of altostratus precipitans and nimbostratus, do not generally produce natural precipitation. As is to be expected, greater degrees of supercooling resulted in a higher proportion of successes. The depth of the supercooled region of the cloud is also important; if this exceeds 4,000 feet, the chances for causing precipitation are very good. Likewise cloud top temperatures of less than −12°C. indicate a high probability of success. In most cases, when precipitation is obtained from a cloud, the cloud dissipates. However, there have been some notable exceptions, when the clouds grew to great heights.

In the supercooled stratiform clouds, there has not usually been very widespread dissipation. However, an interesting case of modification of such cloud occurred during the experiments in which the author participated. There was a continuous layer of stratocumulus, based at 5,000 feet and topped at 5,900 feet. The temperature at the base was 16°F. and that at the top 15°F. The cloud was observed to be composed of liquid water in a supercooled state. The presence of liquid water was indicated by the occurrence of a glory on the cloud top, and by formation of light rime on the aircraft during ascent through the cloud. The cloud was inoculated with solid carbon dioxide from an altitude 400 feet above its top. At this height the temperature was 20°F. The seeding was at a very high rate; approximately 300 pounds of solid carbon dioxide was dropped over a flight path of 8 miles. Another seeding was performed in a similar fashion, approximately 400 pounds of solid carbon dioxide being dropped over a flight path of 10 miles. A half-hour after the seeding, a very striking cloud modification was observed. The area affected was approximately 8 miles long and 3 miles wide (dimensions were determined by making timed flights across the length and breadth of the area). The length of the affected area corresponds with the length of the seeding run, while the width is enormously greater. The rate of lateral spreading of the effect of seeding corresponds very well with the 2 metres per second mentioned by Dr. Schaefer (at a meeting of the Royal Meteorological Society in Toronto). The modification of the cloud, as first observed, took the form of complete disappearance of the roll structure, and its replacement by a depressed area having a fibrous nature and without relief or structural details. As the aircraft approached the seeded area, a pillar of light was observed on the top of the depressed region, opposite the sun. This sun pillar provided evidence that the elements of the affected portion of the cloud had been changed from water droplets to ice crystals. A descent was made through the modified portion of the cloud, and during the descent falling snow was observed, and snow static was heard on the radio. The depression was observed to become gradually an opening in the cloud, with the supercooled droplets all changed to ice crystals, which in turn became falling snow. The edges of the opening were quite well defined, with a narrow zone of downward-streaming snow. The snow was light and fine. The second seeded area appeared to be similar, but darkness prevented detailed examination.

Clouds have occasionally been produced by seeding in damp, clear air. This was done on several occasions during the Arnprior experiments, and also during the author's experiments at Kapuskasing. The clouds so formed at Arnprior tended to persist and become denser with time; those seen by the author were exceedingly tenuous, but were observed for a short time only. The composition of those at Arnprior was observed to be ice crystals, which took the form of vertical curtains, sometimes flattening when formed below a temperature inversion. The manner of formation of such a cloud may well be exactly similar to the process by which seeding produces ice crystals in supercooled water clouds, and is therefore of considerable interest.

In some cases, Langmuir and Schaefer have adopted a very interesting device to demonstrate the ability of seeding to produce openings in clouds. This device consists of seeding in some unusual form, to produce an opening of a shape very unlikely in nature. For example, they have used the form of the Greek letter *gamma*.

The full significance of rainmaking processes cannot be assessed at the present time. Additional experimentation and further study are required. It appears conclusive that each particular situation must be assessed on its own merits, in order to determine whether or not, in that particular locality, with its own particular geographic and meteorological characteristics, and its own particular economic situation, rainmaking can be significant. The difficulty of any attempt to direct rain to any particular spot appears at the present time to be very great. The lack of control so far observed over the rate of fall of induced precipitation might easily be a serious disadvantage. This would, of course, depend upon the purpose for which the rain was wanted. If a storage dam is to be filled, heavy rain should be satisfactory. On the other hand, if an agricultural crop is to be watered, heavy rain is very likely to be almost useless, or even worse than useless, because it may run off too quickly, and may even beat down and destroy the plants. If the induced rain should be accompanied by accidentally induced hail, the attempt to aid the growing crops might well be disastrous. If rain is needed to dampen forest fuels and so reduce fire hazard, a slow rate of fall over a long period is desirable; on the other hand, if it is intended to make an attempt to put out a fire already burning, a high rate of fall may well be needed.

At the present time, it appears certain that the effects of cloud inoculation are strictly local, and do not extend to modification of the air mass. Results, if any, depend upon the characteristics of the air mass concerned. At the time of writing, self-sustaining storms do not appear likely to be caused by interference with natural processes.

7. The Observation of Clouds

Accurate reports of cloud conditions are of the greatest importance not only to meteorologists but to aviation personnel and others engaged in occupations upon which weather exerts a direct influence. Since the majority of cloud observations are entirely non-instrumental, a high degree of skill acquired through much study and practice is required to identify clouds and to determine their heights, the amount of each cloud type and its direction of motion.

IDENTIFICATION

The difficulty involved in correctly identifying clouds varies considerably; some are easily placed in the proper group, while others are difficult border-line cases. The amount of light available very greatly influences the degree of difficulty experienced in this respect. On a moonless night, it is most difficult to determine accurately what types of cloud are present. Lower layers of cloud frequently make observation of middle and high cloud difficult or impossible.

The ice crystal forms of cloud present relatively little difficulty in daylight, although fine distinctions between extensive rather dense masses of cirrus and layers of cirrostratus are not easy. When cirriform filaments are much interlaced, it is a delicate problem to decide whether the layer so formed is cirrus or cirrostratus. When the sun is low, cirrus presents at times a deceptively dark appearance.

Isolated wisps of snow, occasionally seen against the blue sky, present an appearance remarkably like that of cirrus. Close examination reveals that they do not have the pure white colour nor the silky texture of cirrus. Wisps of rain are often thin enough to resemble cirrus, but they are not hard to distinguish, since they are grey and may have rainbows.

The presence of a halo is generally accepted as proof that the cloud is cirrostratus. This criterion is not entirely satisfactory, since rather faint halos appear on rare occasions in thin altostratus. Halos also occur occasionally in cirrus. At times, cirrostratus is so thin that the occurrence of a halo is the only way in which its presence is revealed.

Cirrostratus layers are sometimes difficult to detect at night, because they are so thin that the moon and stars are not appreciably dimmed. The presence of such a layer is sometimes revealed by a lunar halo. A faint whitish aurora may occasionally be mistaken for a patch of cirrus or cirrostratus, but careful examination will detect the motion and illumination typical of the aurora.

The most difficult distinction in connection with these high level clouds is that between cirrocumulus and high delicate altocumulus. The international definition requires association with cirrus or cirrostratus in order that the cloud be properly classed as cirrocumulus. This convention rules out certain patches of cloud which occur alone in the sky, with no visible connection with cirrus or cirrostratus, but having very definitely all the attributes of cirrocumulus. However, the justification of the convention is that it provides a means of eliminating almost entirely the incorrect classification of thin edges of alto-cumulus sheets as cirrocumulus. Close examination of these thin edges shows their connection with altocumulus.

Middle clouds occur in a wide variety of forms, and their identification presents a number of problems. It is difficult to determine when thickening cirrostratus becomes altostratus, and when altostratus becomes nimbostratus. When cirrostratus becomes altostratus the outline of the sun or moon is dimmed, and the cloud loses its pure white colour. The colour distinction will not be observable at night, but the thickening will dim the outline of the moon. If there is no moon the distinction is more difficult, since the stars are often concealed by the thicker forms of cirrostratus.

The different types of altostratus must also be distinguished. Altostratus opacus shows clearly by its appearance that it is lower and thicker than alto-stratus translucidus, and halos would definitely not be expected in the thicker forms of altostratus. The bases of both these forms present a smooth appearance. Altostratus translucidus and thin spots in altostratus opacus have a fibrous appearance in the vicinity of the sun. Thickening of altostratus opacus produces altostratus precipitans, which has a more indistinct base, on account of virga.

Sometimes there is a strong resemblance between altostratus and some stratus clouds. If the cloud is thin enough, the fibrous appearance of altostratus near the sun will provide identification.

The transition from altostratus precipitans to nimbostratus is difficult to detect, because the base of the altostratus is often hidden when this change occurs. It may be hidden by a layer of stratus or fractostratus, or by precipitation. An increase in the intensity of the precipitation is a fairly reliable criterion. An aeroplane observation above the lower layer of stratus or fractostratus is a valuable means of determining the height and nature of the base of the precipitating cloud. The degree of darkening of the sky is not a good indication of this change, since nimbostratus has often the appearance of being illuminated from inside.

Layers of cloud possessing the characteristics of both altocumulus and altostratus are common. Sometimes they resemble one form more than the other. Precipitation from such clouds is quite usual, but if the relief features of the base disappear on account of virga or precipitation, the cloud has become altostratus precipitans.

The distinction between altocumulus and stratocumulus is often difficult, for there are many borderline cases. These forms are in several respects essentially the same cloud, occurring at different levels and having differences in structure and in the forms of water of which they are composed. The differences in appearance are partly the result of the different distances of the clouds from the observer on the ground. The differences in the forms of water are largely the result of the differences of altitude, with consequent differences in the temperature of the environment. A generally accepted rule for distinguishing lower forms of altocumulus from higher forms of stratocumulus is that if the smallest of the regularly arranged elements is less than ten solar diameters in width, the cloud is altocumulus. Ten solar diameters may be estimated by applying the rule that it is the angle subtended at the eye by three fingers at arm's length.

In borderline cases, the seasonal variations in the altitudes of these cloud forms must also be considered. The upper limit for stratocumulus may be expected in summer to be of the order of 7,000 to 9,000 feet, and in winter 6,000 to 7,000 feet. These limits are for middle latitudes. The heights mentioned will be appreciably exceeded in the vicinity of high mountains.

There is naturally considerable difference in the apparent coarseness of the cloud structure when the cloud is viewed from different altitudes. A cloud which would appear to an observer on the ground as altocumulus at 10,000 feet would appear to an observer in an aircraft at 5,000 feet as stratocumulus at 5,000 feet.

Difficulty is also often experienced in correctly identifying the low cloud forms. In particular, there are many cases in which other low cloud forms are likely to be confused with stratocumulus. Layers of stratocumulus composed of fairly regularly arranged elements with blue sky or thin spots between are very easily identified as stratocumulus. Stratus with large loops hanging from the base is rather easily confused with stratocumulus, but the extreme coarseness of its structure shows it to be stratus. A layer of fractostratus with its separate elements quite close together may also easily be mistakenly reported as stratocumulus. Such a layer of fractostratus is often a stage in the formation of stratus. Stratus occasionally has some relief in its base, but if that relief acquires the form of regular rolls, the cloud becomes stratocumulus. Observations from aircraft of the top of the cloud layer will often reveal the nature of its structure. Fractostratus (sometimes merging to form stratus) and stratocumulus form under very similar conditions, at the top of a mixed layer of air. There are many borderline cases, where great care is necessary to identify the cloud correctly.

Large numbers of separate cumuli, when viewed from a distance, have the appearance of stratocumulus. This is the effect of perspective.

Some difficulty is occasionally experienced in distinguishing fractostratus from stratus. Once the cloud has become continuous, with no breaks or only a few breaks, it is stratus. There is sufficient difference between fractostratus and fractocumulus that no difficulty should be experienced in distinguishing them. Fractocumulus shows generally that it is the forerunner of cumulus, having the form of little balls of cloud or of elongated wisps. The form assumed depends largely upon wind conditions.

Nimbostratus presents the observer with certain difficulties. Its appearance is much like that of stratus. However, if steady precipitation is occurring in the form of raindrops or snowflakes or combinations of these, the cloud is nimbostratus.

Accurate observation of the base of nimbostratus is difficult and often impossible. Precipitation in the form of rain falling from nimbostratus causes the cloud base to build downward, forming a layer of stratus, frequently based at a very low level, and merged at its top with the nimbostratus. Such a layer of stratus becomes virtually a part of the nimbostratus. In other cases, under the influence of mixing, combined with the addition of water vapour to the air, a layer or patches of stratus or fractostratus will form under the nimbostratus and separated from it, obscuring the base of the nimbostratus from the ground observer. When the precipitation is in the form of snow, the base of the nimbostratus is concealed by the precipitation itself. When observable, the base of nimbostratus is soft and indefinite. As mentioned, occurrence of continuous precipitation, other than drizzle, from a low cloud is adequate evidence that the cloud is nimbostratus.

Observation of the clouds of vertical development involves decisions regarding the transitions from cumulus humilis to cumulus congestus and from cumulus congestus to cumulonimbus. Accurate observations of these changes are of very great importance to the forecaster, since the extent of vertical development and the forms of the tops of these clouds provide most valuable indications of probable shower or thunderstorm activity.

Cumulus clouds present a characteristic appearance, with flat bases and rounded tops. These clouds occasionally increase in amount until they cover the whole sky, merging to form a layer of stratocumulus. Just when the change to stratocumulus becomes complete is sometimes difficult to judge, but the cloud has not become stratocumulus until the bases of the separate cumuli are fused into one layer of cloud.

The transition from cumulus humilis to cumulus congestus involves upward growth of the cloud tops and, frequently, heavy swellings. Cumulus congestus has a very active aspect, with "boiling" tops. Towers, with fairly narrow cross-sections often develop. The cloud has hard outlines, which frequently reflect sunlight strongly, and produce a brilliant appearance about its top.

The change from cumulus congestus to cumulonimbus takes place when the top of the cloud becomes glaciated. The hard outlines become softened, and cirriform fibres appear. The cirriform portions of the top may or may not spread out to form an anvil. Cumulonimbus calvus or "bald cumulonimbus" has no anvil.

In some situations the tops of clouds of vertical development are concealed from the observer on the ground by a layer of lower cloud, either below their bases or containing their bases. In other cases, these clouds penetrate a layer cloud from below. Cumulonimbus is also often concealed by large numbers of cumulus clouds. The occurrence of showers of moderate or heavy intensity constitutes sufficient evidence to justify the assumption that cumulonimbus clouds are present. Occurrence of hail or of thunder and lightning is positive proof of the presence of cumulonimbus.

A deceptive appearance is sometimes presented by the anvil top of a distant cumulonimbus incus with cumulus congestus between it and the observer. This is sometimes erroneously reported as cumulus congestus and cirrostratus. The lower portions of the cumulonimbus are obscured by the cumulus congestus, or are below the horizon.

AMOUNT OF CLOUD

Cloud amounts are expressed in terms of the fraction of the sky covered. For some purposes, tenths of sky are used, and for others, eighths. This part of the observation is almost always entirely non-instrumental. From an observation point commanding an unobstructed view of the sky, the amount of cloud may be determined by estimation. It is found satisfactory to face successively in two opposite directions, so dividing the sky in halves. The amounts of the several cloud forms in each half may then be estimated, and the results added together. The halves of sky may be chosen in such a way as to simplify the estimation. For example, if the cloud is all in the western half of the sky, a division into east half and west half is indicated. Effects of perspective increase the difficulty of estimating cloud amounts near the horizon. It is useful to remember that half the sky is above a 30° angle from the horizon.

A rather uncommon device for measuring the amount of cloud consists of a hemispherical mirror with divisions marked on it. The mirror is mounted where there is an unobstructed view of the sky, and the number of divisions covered by the reflections of the clouds is determined by counting. Another device sometimes used consists of a large hemispherical framework of metal rods, which form a grid. The device is of a suitable size and is suitably mounted for the observer to stand under it, and count the number of divisions of the grid where clouds are observed.

The *polestar recorder* is a rarely used device for determining cloud amount at night. It consists of a very long focus camera, which is directed at the north pole of the sky. The plate is exposed throughout the night. The path of the star Polaris will be recorded on the plate, with breaks in the record caused by clouds. Measuring the proportion of the path obscured furnishes an estimation of the cloud amount. To be successful, this method depends upon a uniform distribution of cloud over the sky. Obviously, then, its usefulness is not great.

HEIGHT OF CLOUD

The heights of clouds are determined by several different methods, which may be roughly classed as *estimation* and *measurement*.

The method of estimation is applicable in daylight and in bright moonlight. Under suitable light conditions an experienced and careful observer obtains very satisfactory results in determining cloud height in this manner. A visual estimation is made of the distance from the ground to the cloud base. In making this estimation, the observer studies the cloud structure. When the structure of the cloud is delicate, there is a tendency to overestimate its height. As low clouds affect the safety of flight more directly than do the higher forms, their heights must be estimated with greater care. Estimation of the heights of clouds is often easier if an aircraft is seen at or very near the base of the cloud. The height of the aircraft may sometimes be estimated more readily than that of the cloud. In any event, it provides a useful aid in estimating cloud height.

Estimation of the height of the cirriform clouds is rather difficult, but may be accomplished by an experienced observer. Cirrocumulus, on account of its form, is the cirriform cloud best adapted to height estimation.

The heights of layers of altostratus are rather difficult to estimate, on account of the lack of relief in their bases. The thinner forms, however, are known to be higher than the thicker forms. When altostratus thickens to altostratus precipitans estimation of its height becomes more difficult. Virga trailing from the base of the cloud cause the base to assume an indefinite appearance. This difficulty is further increased by the formation, after rain commences, of layers or patches of fractostratus, stratus, or stratocumulus. Altocumulus, on account of its form, lends itself admirably to height estimation. The relief on the base of layers of combined altocumulus and altostratus is sufficient to make such a layer fairly simple to estimate.

There is considerable variation in the degree of difficulty experienced in estimating the heights of the low cloud forms. Those having relief features on the base are much more satisfactory from this point of view than those having uniform flat bases. Stratocumulus, like altocumulus, is very well suited to this method of height determination. It must be remembered, in this connection, that stratocumulus presents a great variety of appearances and that a finer structure than usual may cause difficulties. Estimation with any degree of accuracy of the height of stratus layers is often impossible, although a rough value may be obtained by an experienced observer. Fractostratus, on account of its broken character, lends itself to this method quite satisfactorily. Estimation of the height of nimbostratus is in most cases virtually impossible. On the rare occasions when it is present without precipitation actually falling, its base is indefinite on account of the presence of virga. When precipitation is occurring, the precipitation itself and also low cloud formed in the precipitation make observation of the base of the nimbostratus difficult or impossible.

Fractocumulus has an irregular form, and its height may be estimated with reasonable ease. Cumulus humilis and cumulus congestus are not particularly well adapted to this method, but a degree of accuracy sufficient for practical purposes may be obtained. The bases of cumulonimbus clouds are often ragged and indefinite, with fractostratus below them, and so determination of their height is difficult. This difficulty is increased by the heavy precipitation below the cloud. The snow squalls associated with winter cumulonimbus completely obscure the cloud base.

Formulae have been developed for determining approximately the height of clouds in situations when the air below the cloud base is thoroughly mixed. These formulae make use of the surface temperature and dew point. This method may be also considered a means of forecasting cloud heights. It is best adapted to thermal cumulus.

The temperature and dew point of the air both fall with height, the temperature falling more rapidly than the dew point. Hann-Suring published a formula resulting from a comparison of the calculated rates of change of these two quantities:

$$Z = 220 \, (T - T_d) \text{ feet}$$

Where T is the surface temperature and T_d the surface dew point in degrees Fahrenheit, and Z the height in feet of the cloud base above the surface. Psychrometers with built-in nomograms for using the relation between temperature and dew point and cloud height are produced by several makers. The formula must be used with caution, and is not at all suitable over mountainous terrain.

A competent observer in an aircraft obtains the most accurate data on the height and nature of the bases of the clouds. If the aircraft is suitably equipped with radio, the tops of the lowest clouds, and also the tops and bases of higher layers may be investigated. A complication is introduced if the cloud layers concerned are at temperatures below freezing, for in this case an aircraft suitably equipped with de-icing equipment is required.

Other methods of measuring cloud heights suffer from the serious defect that the measurement is made in one spot only. In some cases a duplication of equipment makes possible more complete measurement. Most of these other methods also fail to determine the thickness of the clouds and to detect the presence of higher layers.

A rather crude determination of the heights of cloud bases and tops is furnished by the records of radiosondes. This depends upon the relative humidity being indicated as 100 per cent while the balloon is in the cloud. A quite accurate determination may be made if the radiosonde balloon is seen to enter the cloud base, and the balloon's height at the time of disappearance is determined from the pressure signals.

Small balloons known as *ceiling balloons* are designed for the purpose of measuring the height of the cloud base. Those in use in Canada are red in colour. The balloon is filled with hydrogen or helium until it has sufficient free lift to ascend at a predetermined rate (usually 140 to 150 metres per minute). It is then released, and the time required for it to reach the cloud base is determined. The rate of ascent being known, the height of the cloud base is easily computed. This method, while very neat and simple, is far from perfect. The height of the cloud base is measured in one spot only, that at which the balloon enters the cloud. If there are sizeable breaks in the cloud, the balloon often ascends through these breaks. In strong gusty winds the balloon will not produce satisfactory results, particularly if released from the roof of a building. In such cases, the downdraft on the lee side of the building is very likely to carry the balloon to the ground, where it travels for a short time at ground level before its ascent begins. This makes the timing very difficult and uncertain. If there is rain falling at the time of the observation, the weight of the rain on the balloon tends to slow its ascent, and so to cause an inaccurate height determination. Overfilling the balloon will to some extent correct for this effect, but is dangerously inaccurate. Even in the absence of disturbing factors, the rate of ascent of a ceiling balloon is very erratic.

There is an unfortunate tendency for red ceiling balloons to show much too clearly after they have actually entered the cloud base, and so to give an excessively high value for the height of the cloud. This is notably the case with very low and ragged clouds, where the actual distance from the observer is small, and the lower portions of the cloud are rather tenuous. It also sometimes happens that the balloon enters the cloud at a spot where the base is higher than its general altitude, thus giving an excessively high value. An interesting example of this particular defect was provided by an occasion when the writer was in an aircraft which entered the cloud base at 100 feet, while a balloon released within a very few minutes of the same time at the same place was quite clearly visible to an observer until it reached a height of 200 feet. As soon as the balloon begins to fade at all it must be considered to be in the cloud. Use of this method at night would require that a light be carried on the balloon. To lift such a light requires a balloon larger than a ceiling balloon.

Pilot balloons provide cloud base determinations in addition to their principal function of measuring upper winds. At night, lighted pilot balloons provide the same information.

A much-used method of measuring cloud heights at night involves the use of a device known in Canada and the United States as a *ceiling projector* and in the British Isles as a *cloud searchlight*. A searchlight projects a powerful narrow beam of light on the base of the cloud. On moonless nights well-defined spots of light are obtained on clouds as high as 15,000 feet.

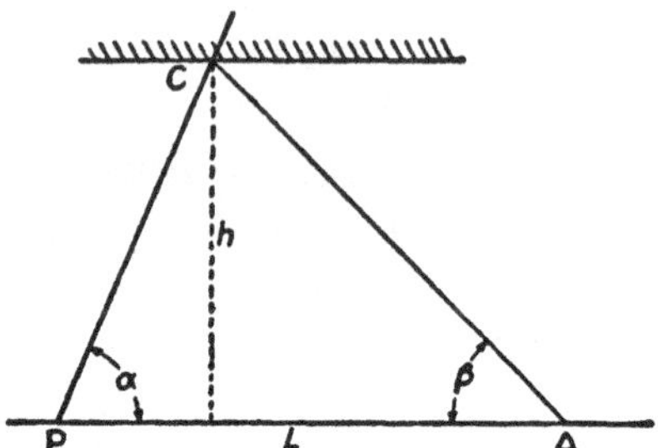

FIG. 5. The theory of the ceiling projector. (From *Meteorological Instruments*, by W. E. K. Middleton, published by University of Toronto Press, 1947.)

The principle of the ceiling projector involves the application of simple trigonometry. After Middleton, let Z be the cloud height, L the length of the base line, α the projection angle and β the angular elevation of the spot of light.

Then: $L = Z \cot \alpha + Z \cot \beta$ and the cloud height is:

$$Z = \frac{L}{(\cot \alpha + \cot \beta)}$$

If the beam is projected vertically $\cot \alpha = 0$ and $Z = L \tan \beta$.

Complications are introduced if the projector and the device used for measuring the elevation of the spot are not at the same height. These difficulties are easily resolved by a suitable alteration in the length of the base line.

The angular elevation of the spot is measured by means of an *alidade* or a *clinometer*. Some alidades have scales graduated to read cloud heights directly. Some projectors are set to produce a vertical beam, and others to project a beam at an angle of 71° 34'. The vertical beam provides more accurate determination of the higher cloud bases. The 71° 34' projection angle was chosen with a view to obtaining maximum accuracy for cloud bases in the vicinity of 3,000 feet. With this projection angle the spot of light is directly overhead when the cloud base is 3,000 feet and the base line is 1,000 feet (71° 34' is $\tan^{-1} 3$).

By means of the projector, cloud heights may be determined at night with a very satisfactory degree of accuracy. However, ragged cloud bases present a problem, as the spot of light is elongated and not sharp. In such cases, the observer sights on the lowermost portion of the elongated spot. Unfortunately, the projector measures the height of the cloud base in one spot only. By using two or more projector installations at the same observing station, height determinations may be made on more than one spot on the cloud base, and greater accuracy attained. Careful use of the projector will often make possible measurement of the heights of several layers of a broken or scattered cloud. In such cases, more than one spot may be visible at the same time.

Because the brightness of the sky is so much greater than that of the spot of light from the projector, this method cannot be used in the daytime. A device for determining cloud heights in the daytime has been developed by the National Bureau of Standards of the United States, following a suggestion by Middleton. A modulated beam of light is employed, which is identifiable on account of its difference from the steady light of the sky. The projector of this device employs an A.C. operated mercury arc as a source of modulated light. The "ceilometer," which is the measuring instrument, consists essentially of a photoelectric detector and an indicator to show the angular elevation of the spot of modulated light on the cloud base.

The height of the base of a cloud may be determined by two observers with theodolites. The method is similar to that employed in two theodolite pilot balloon observations. As a means of measuring cloud height this method has some very serious defects. Both observers must sight simultaneously on the same detail of the cloud base. For two observers to identify the same detail from their different observation posts is very difficult, on account of the pronounced difference in the appearance of the cloud base from different directions. There is also the manifest difficulty of describing over a telephone the particular detail in such a way that both observers can recognize it. Further, identifiable cloud features are subject to constant change. This method, in view of the defects mentioned, is excessively extravagant from the point of view of personnel and equipment required. The difficulties may be overcome by the use of two photo-theodolites making simultaneous photographs. However, the time consumed by this method restricts its usefulness. It is helpful for research purposes, but is not a suitable technique for ordinary observations.

An interesting and ingenious single theodolite method may be used in certain very special circumstances. It requires that the observer be in a position to set up his theodolite on a hill or tower so located that he may observe the angle of elevation of a feature of the cloud base and also the angle of depression of its image reflected in a pond at the foot of the hill or tower.

A theodolite may be used in conjunction with a range finder to measure the height of a cloud with a sharply defined base.

An *optical radar* device has been employed to determine cloud heights. This device depends upon a measurement of the time required for a light beam to reach the cloud base and return.

Range finders are occasionally used to measure cloud heights. The writer has been informed by Dr. C. F. Brooks that at the Blue Hill Observatory it has been noted that only some observers have eyes suited to the use of range finders.

DIRECTION OF MOTION

The direction and speed of a cloud may be determined with the aid of a *nephoscope*. These instruments are of two general types; in using one type the observer looks directly at the cloud through the nephoscope, and in using the other he looks at a reflection of the cloud.

The *comb* nephoscope consists essentially of a series of rods assembled in the form of a comb, and mounted on a vertical rod with the "teeth" of the comb horizontal. The vertical rod is free to turn, and is rotated by means of a pulley and rope device. The observer selects a recognizable feature of the cloud base, and then rotates the comb of the nephoscope until that feature appears to move along the longitudinal member of the comb. To aid in keeping the eye trained on the cloud and in determining its direction correctly, a sharpened stick may be used; the stick is set in the ground and a sight is taken from the point of the stick to the comb to the cloud. The direction of the cloud's motion is read off from a pointer connected to the vertical rod and travelling over a horizontal circle marked in degrees. By using the "teeth" of the comb, and a timepiece, the observer may time the motion of the cloud. The speed of the cloud may then be determined, provided its height is known.

The *grid* nephoscope is a later modification of the comb nephoscope. In this instrument, the comb is replaced by a grid of rods in two directions at right angles to each other. The method of operation is essentially the same; the device being rotated in such a way that one of the longitudinal members of the grid is kept in line with the cloud. The cross-pieces are used for timing the motion of the cloud in the same manner as the teeth of the comb nephoscope.

The *mirror* nephoscope consists essentially of a horizontal disk of black glass engraved with concentric circles. A vertical pointer adjustable to various known heights is mounted on the rim of the instrument. The pointer is so placed by the observer that the image of the cloud under observation is in line with the pointer and the centre of the disk. The image is kept in line with the pointer as it moves, and the direction of motion is read off from a scale on the rim of the circle.

Another type of mirror nephoscope has a mirror on which the only marking is a centre point. An eyepiece, which may be adjusted to suit observation of clouds at different levels, is provided. The image of the cloud is observed through the eyepiece, and followed from the centre of the mirror with a small triangular metal marker. A ruler is then laid on the mirror through the centre point, the marker point, and the azimuth circle. The distance from centre point to marker with the eyepiece height is used to compute the speed of the cloud's motion.

Prior to the extensive use of pilot balloons, nephoscope observations were widely used to determine the speed and direction of the upper winds by determining the speed and direction of the clouds. In order that wind speeds may be computed by this method, the heights of the clouds must be known with a reasonable degree of accuracy. Nephoscopes are at present very little used. The direction and speed of a cloud is best determined by observing the direction and speed of a pilot balloon at the time it enters the cloud. If the height of clouds for use in wind determinations is to be measured by balloon, it is obviously much better to use the balloon directly as a means of determining the winds at various levels. Further, the balloon will find the winds at levels where there are no clouds. If the cloud height used in determining wind speed is an estimated height, the speed determination becomes rather inaccurate.

The direction of motion of a cloud may be determined by successive observations with a theodolite of some recognizable feature of the base. The theodo-

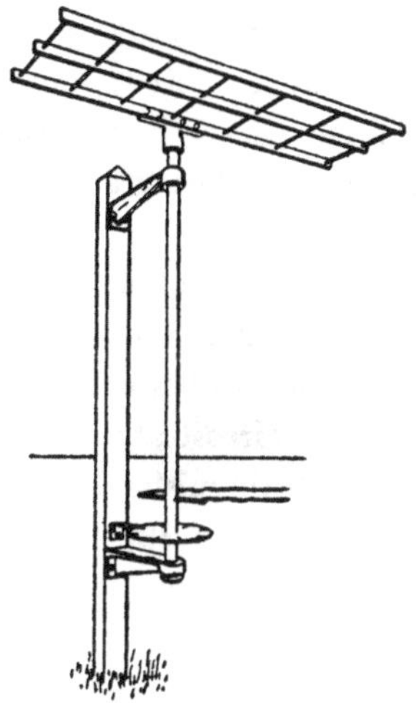

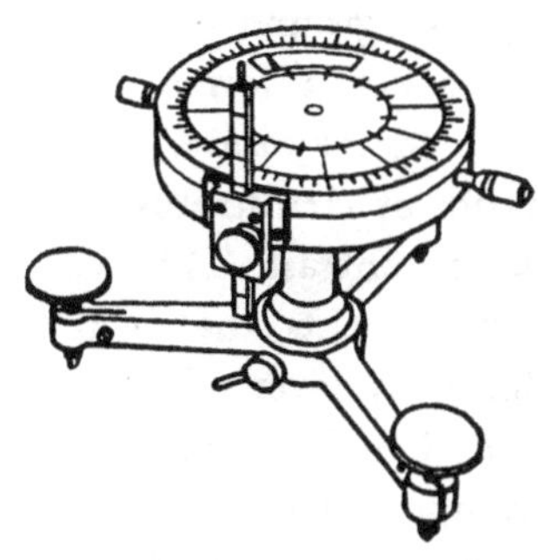

FIG. 6. Grid nephoscope. (From *Meteorological Instruments*, by W. E. K. Middleton, published by University of Toronto Press, 1947.)

FIG. 7. Mirror nephoscope. (From *Meteorological Instruments*, by W. E. K. Middleton, published by University of Toronto Press, 1947.)

lite used for this purpose need have sights only, since a telescope is not necessary for the distances involved.

The speed of motion of the cloud may be determined by timing in conjunction with theodolite observations. This requires knowledge from some other observation of the height of the cloud. If this other observation is a balloon observation the same argument as used in connection with the nephoscope will apply; it is better to use the balloon directly as a means of determining the speed and direction of the wind. If the height determination is an estimate, another approximation is introduced, and the accuracy of the wind determination becomes less than ever.

CLOUD PHOTOGRAPHY

Photography provides an excellent means of recording the appearance of cloud forms. This is a specialized form of photography, and presents several problems not generally encountered in other types of photographic work.

Certain cloud forms are definitely photogenic, while others are very difficult or impossible to photograph with any degree of success. In general, the forms which present good contrasts are, as is to be expected, the easiest to photograph. Cumuliform clouds in general, and in particular the bulky forms of the clouds of vertical development, are the most satisfactory from the photographer's point of view. The cumuliform layer clouds, if they have interstices between elements, and no higher layer of cloud is present above, present very satisfactory contrasts. Non-continuous forms always present much better contrasts than continuous forms. The undulated clouds with no interstices present much greater difficulties, and the veil clouds are very difficult, and, indeed, often impossible to photograph with any degree of success. A cloud form which photographs satisfactorily against a background of blue sky will often not photograph at all well against a background of higher cloud.

The high cloud forms require care on the part of the photographer, particularly in the exposure. Cirrus and cirrocumulus photograph well, while good photographs of cirrostratus are somewhat more difficult to obtain on account of the lack of contrast. This will not apply so much to cirrostratus composed of interlaced filaments, that is, the form which is barely distinguishable from cirrus. The edge of a layer of cirrostratus not covering the entire sky provides good contrast and such a cloud will show in a photograph to better advantage than will a complete layer.

Altocumulus in many of its forms is photogenic, since there are very commonly interstices between elements, providing good contrast. Layers of altostratus, on the other hand, are not nearly so easy to photograph, on account of the lack of detail. Occasionally, if the sun is in the field of the picture, a good representation of the cloud may be obtained. The lenticular forms of altocumulus photograph well in many cases, but require great care with the exposure.

Certain forms of stratocumulus are photogenic. Those forms having elements separated by clear spaces may be satisfactorily photographed, in much the same way as the similar forms of altocumulus. Roll stratocumulus, on the other hand, is generally so dense that the contrast is not satisfactory and the relief frequently fails to show to advantage. Even the more easily photographed forms of stratocumulus often do not provide satisfactory results if photographed against a background of higher cloud. The contrast is not good, and the illumination is inclined to be poor. Fractostratus, against a clear sky, is satisfactory if it is scattered or broken. However, if it forms an overcast with only a few breaks it is difficult to work with for the same reason as the roll stratocumulus. Against a background of higher cloud it will generally provide rather poor results, unless the higher cloud is thin enough or broken enough to let through considerable light. Against a low sun fractostratus shows well in a photograph. The veil clouds, nimbostratus and stratus, are not photogenic. Nimbostratus, in particular, is generally impossible to photograph. In almost all cases, nimbostratus is producing precipitation, which hides the cloud to some extent if it is rain and completely if it is snow. In addition to this difficulty, the rain or snow is quite sure to get on the lens of the camera, or on the window if the picture is made from inside. A photograph taken from under a roof with no window between the camera and sky will obviate this difficulty. Stratus seldom produces a photograph which is of any use at all. However, against a background furnished by a sufficiently high hill it sometimes shows reasonably well. An excellent photograph of stratus has also been made with the Washington Monument fading into the cloud.

The clouds of vertical development furnish the photographer with most satisfactory subjects. Contrast is generally excellent, the clouds being in most cases sufficiently non-continuous. The presence of a layer of cloud above greatly reduces the chance of satisfactory results. Rather striking photographs may be obtained when the sun is low, with a cumulus cloud between it and the photographer. The crepuscular rays produce very interesting and artistic effects.

Mammilated forms of cloud generally occur under very difficult lighting conditions. Although the relief features are of very considerable extent, the unsatisfactory lighting and the relatively small amount of contrast available make this a difficult cloud to photograph with any conspicuous degree of success. The very great relief features, however, provide better light and shadow effects than are usually obtainable with roll stratocumulus.

Cloud photography is greatly facilitated by suitable filters. By the use of such equipment, the photographer is able to reduce the intensity of the highly actinic short wavelengths of light and so to give the other colours a better chance to register on the film. The choice of filters will be governed by the purpose for which the photograph is required. For purposes of record of the cloud's features and details and for studies of the cloud's structure, it is desirable to reduce the intensity of the short wavelengths of light entering the camera, but not so much as to sacrifice detail for contrast. On the other hand, if an artistic effect only is desired, or if contrast at the expense of detail is desirable, a denser filter may be used. The writer finds a deep yellow filter of the type commonly known as "Wratten G", very satisfactory. A lighter yellow filter was found insufficiently dense. For greater contrast, with some sacrifice of detail, a red filter such as the "Wratten A" may be used. The use of a red filter restricts the choice of film, as panchromatic types must be employed in order that any image at all may be obtained.

The colour of the sky is an important consideration. Better contrast may be obtained with white clouds against a bright blue sky than with white clouds against a pale blue sky. This difficulty of photographing against a pale sky is increased when the clouds concerned are high and thin.

The choice of film merits careful thought. Since in cloud photography a faithful recording of detail is important, especially if the photograph is to be enlarged, a fine grain film is more desirable than a faster but coarser grain film. There is generally more than ample light available, and so nothing is gained by using a fast film, and fine detail is not so well recorded on some of the faster films. As already mentioned, if a red filter is to be used, choice of film is restricted to the panchromatic types. To preserve the detail it is advisable to employ a fine grain developer. Care in printing and enlarging will assist further in bringing out the maximum possible detail.

In work of this kind, as in all photographic work, the best lens available is the most satisfactory, since detail is more faithfully recorded, and negatives permitting greater enlargement are obtained. A fast lens is not required, since it is usually very considerably stopped down, but the better definition provided by the more highly corrected lenses is very desirable. In view of the distances from the photographer at which the clouds are generally located, focussing does not present a problem. Further, at the small lens openings employed, there is considerable depth of focus. On account of the magnitude of the distance, smaller lens openings with consequent better definition are practicable, since longer exposures do not generally result in any loss of detail; the cloud is far enough away that its apparent motion is not in general sufficiently rapid to cause any blur. The aid of an exposure meter is most valuable, since it makes possible more accurate determinations of exposure and lens opening. Under-exposure or over-exposure may result in the loss of an opportunity to obtain a photograph of a very rare cloud condition.

Photographs of clouds from above provide very interesting effects, and show details which assist in classification. Lighting is generally considerably better than below the clouds. Lower broken or scattered clouds usually show quite well against a background of the earth; not so well if the earth is snow covered. In general, rather better photographs of fractostratus may be obtained from above than from below.

Very interesting effects are obtained by the use of wide angle lenses, for the whole sky may be included in one photograph. A technique occasionally employed in photographing the whole sky involves the use of a hemispherical mirror. The mirror is arranged in such a way that it reflects the sky, and a camera is used to photograph the reflection in the mirror.

Photographic methods have been used successfully in determining the heights of the stratospheric cloud forms. The same technique is used as in determining the height of the aurora. The cloud is photographed simultaneously against a background of stars, from two points of observation a known distance apart.

RADAR AND CLOUDS

Certain meteorological phenomena give rise to radar echoes. The return from an aggregation of water drops is proportional to "Z," where Z is the sum of the sixth powers of the drop diameters. It has been found that echoes are caused by both rain and snow. They are also produced by parts of some clouds.

A horizontal "bright band" has been observed in clouds. This band has been detected with the aid of height-finding radar equipment. When such a band occurs, it is found to be at, or near, the freezing level. It has been ascribed to an increased reflecting power, caused by a coating of liquid water on melting ice crystals. Such water-coated ice crystals would have a larger effective reflecting area than would either water droplets or ice crystals alone.

Cloud droplets of the more common sizes do not cause radar echoes with the types of 10-cm. and 3-cm. radar equipment at present in use. However, when raindrops are present in and below the cloud, echoes are produced. Radar of 10-cm. wavelength will generally not detect rain of less than a certain minimum intensity. An investigation performed by the author has indicated that the minimum detectable rain, at a range of 20 miles, is approximately one-thirtieth inch per hour.

Radar is thus a useful additional observing instrument. It has been found to be of particular value in studying the motions of hurricanes.

8. Clouds in Relation to Forecasting

IT IS convenient to divide the consideration of clouds in relation to forecasting into two separate but closely related sections. A study of prevailing cloud conditions, both as observed locally and as reported from distant stations, provides the forecaster with much useful data. When correlated with other information available from synoptic and upper air charts, cloud data make possible a more accurate analysis of the current weather situation. A more accurate analysis of the meteorological situation prevailing makes possible better forecasts of future weather developments. A most careful study of local cloud conditions is vital to the success of single station forecasting.

On the other hand, prediction of cloud types, amounts, and heights, together with associated weather conditions, presents one of the most difficult and complex problems with which the meteorologist is concerned.

Complete and accurate cloud information indicates the nature and extent of developments taking place in the upper atmosphere and provides useful indirect evidence of moisture content and stability conditions of the air. While part of this upper air information may be obtained by the use of sounding balloons or aircraft ascents, such measurements are of necessity limited both in coverage and in frequency. Even where instrumental observations are available, cloud observations provide much additional information for correlation with the instrumental records and supplement them by giving a continuous indication of developments between the times of soundings.

A study of past observations of clouds furnishes useful information regarding what might be termed the "cloud history" of the air mass or air masses concerned. The history of moisture content and stability and of motions of air is thus provided. By correlating these past conditions in the air masses with present conditions, the meteorologist may determine to what extent the air masses have been modified. This history of modification should assist in estimating future modifications.

The problem of detailed cloud forecasting such as is required for aviation purposes is one of enormous complexity. Here not only the cloud type is important, but also its height, amount, and thickness. Conditions within the cloud with respect to icing and turbulence are also of considerable importance, and must be mentioned in aviation weather forecasts.

Forecasts of cloud conditions must be based not only on large scale air mass motions, but also on small scale local effects. Local effects of very great importance are produced by irregularities of terrain and also by bodies of water. It is, accordingly, of the utmost importance that the meteorologist familiarize himself with the topography of his forecast district.

Observations of high cloud forms supply information regarding the presence of overrunning warm air, its moisture content, and its direction of motion. Thickening of this high cloud indicates an increase in the depth of the air affected by the lifting. A careful watch on high clouds in conjunction with a study of pressure tendencies will often provide most valuable indications of the movement of frontal systems. This is particularly true when changes in direction of the frontal systems occur. By this means skies which appear at first to be front zone skies are often subsequently found to be lateral zone skies. Recession of high cloud, with rising pressure tendencies, will often indicate such a change of direction. Rather small amounts of lenticular clouds at high levels are produced by indirect lifting of air, but these will not be confused with frontal high cloud, or high cloud formed as a result of convergence. Formation of these lenticular clouds indicates the presence of a layer of air with high relative humidity at the level at which the clouds form. Presence of cirrocumulus provides an indication of a small amount of turbulence at high levels.

The exact type and amount of high cloud are not easy to forecast, but there are relatively few activities which are much affected by clouds at these levels. In order to forecast these cloud forms, it is necessary to decide if there will be overrunning warm damp air, and if so how much there will be. If the layer of damp air affected by lifting is expected to be shallow, cirrus only is to be expected; if it is expected that a deeper layer will be affected, cirrostratus should be forecast. Forecasting of cirrocumulus would in general be much too difficult, since the requisite condition of slight turbulence could not be satisfactorily anticipated. This cloud is not important enough in its effects to make such very difficult decisions worth attempting.

Large masses of middle cloud indicate large scale lifting of moist air by overrunning or undercutting or by orographic effects. Thickening of such cloud masses indicates deepening of the layers of air affected. The very important change to precipitating middle cloud is indicated by the actual occurrence of precipitation at some of the stations in the district under consideration. A study of middle cloud systems along with a careful watch on the pressure tendency indicates the direction of motion of the frontal systems producing these clouds. At his own station the forecaster may profitably watch for the transition from altostratus opacus to altostratus precipitans.

Indirect lifting of moist layers of air causes formation of lenticular altocumulus Altocumulus is formed in relatively small quantities by changes in cumulus. in the lee of hills and also above "convective elements." Such clouds indicate the presence of a layer of moist air at the level where they form, and also some disturbance in that layer of moist air. Instability at middle cloud levels is indicated by formation of altocumulus castellatus. Under more stable conditions, layers of altostratus or of mixed altostratus and altocumulus will form. Considerable altocumulus may be formed in quite stable air, with a scheme of motion in the form of cells.

Forecasting of middle cloud constitutes a very complex problem. This problem is rendered more important from the viewpoint of aviation by the fact that precipitating middle clouds are frequently responsible for the formation of low cloud beneath them. In other situations, the presence of a layer of middle cloud above an already formed layer of low cloud will seriously reduce the solar heating available for the dissolution of the low cloud. Middle cloud is also much more often found at or below flight levels than is high cloud.

Forecasting of large masses of middle cloud involves decisions regarding the amount of overrunning or undercutting and the direction of motion of frontal systems. The amount of orographic lifting, if any, is also significant. In all these situations, the moisture content of the lifted air is of the very greatest importance.

Forecasting of the smaller quantities of altocumulus formed from cumulus and also the relatively small amounts of lenticular cloud formed by indirect lifting, is in general of very slight importance.

The presence of low cloud indicates saturated air at low levels, and the form assumed by the cloud gives some indication of air motions. Stratocumulus often indicates air motion in the form of cells, and sometimes appears to indicate motion in the form of waves. Instability at fairly low levels is shown by the presence of stratocumulus castellatus. Low cloud of the fractostratus, stratus, or stratocumulus types often indicates vertical mixing in the lowest layers of the atmosphere. Fractocumulus shows the presence of thermal or orographic updrafts, and is frequently an indication of future formation of cumulus. "Flattening" of the updrafts is indicated by a change in the form of cumulus to produce stratocumulus. Formation of nimbostratus indicates lifting of great masses of moist air.

Because low clouds constitute a hazard to aviation, forecasters serving aviation must exercise very great care in predicting these cloud forms. A number of factors are involved in forecasting low clouds which are not significant in connection with high and middle clouds. Topographical features and bodies of water assume much greater importance. Further, low clouds are quite often producers of precipitation, and their activity is thus of considerable importance.

Nimbostratus is a highly significant form, for it produces large amounts of precipitation and is, furthermore, a very thick and often very low cloud. It may be expected to form when great masses of air of high moisture content are lifted by fronts or by orographic effects.

Stratocumulus is produced by a variety of processes, and may be expected to form when any one of several conditions is anticipated. The spreading of the lower portions of cumulus clouds late in the day, and the merging of cumulus humilis early in the day will both form stratocumulus. Vertical mixing extending through a fairly deep layer of air produces sizeable masses of stratocumulus cloud, notably in fresh Polar Continental air. Addition of water vapour to the air by evaporation of rain falling from above, combined with mixing, will cause formation of considerable quantities of stratocumulus.

Stratus is formed by mixing, by orographic lifting, by lifting of fog, and by addition of water vapour to the air by evaporation of rain. It is formed by mixing in cases where the mixed layer is shallow and there is relatively little turbulence. The first cloud formed in these cases is usually fractostratus, which merges to form stratus. Mixing may be combined with addition of water vapour to the air from evaporating raindrops. The height at which stratus will be formed by mixing may be successfully forecast by considering the surface temperature and dew point, and assuming a dry adiabatic lapse rate throughout the mixed layer. For example, if the surface temperature and dew point are 2°F. apart, and the relative humidity of the air is believed to be high for the first few hundred feet above the surface, stratus may be expected to form at 400 feet, provided there is sufficient wind to ensure thorough mixing in the lowest layers of the air. What constitutes "sufficient wind" will vary with the topography of the region, and is a question which is best answered by acquiring an empirical knowledge of cloud formation conditions in the district concerned. Orographic lifting will produce stratus when the air flows against a hill or hills of suitable height, and the wind is sufficiently strong to force it to ascend to its lifting condensation level.

The forecaster must decide not only whether stratus will form, and at what height, but how long it will persist. The orographic type may be expected to persist so long as the current of air with high relative humidity flows against the hill, or until the wind becomes strong enough, or solar heating becomes sufficient to dissipate the cloud. Stratus formed by mixing will have an endurance depending upon the moisture content of the air, and upon the cloud cover above. When the moisture content of the air is high in the lowest layers only (as often after a clear night), the cloud will not persist for long, since it will be dissipated by solar heating. How rapid and effective this heating will be depends upon whether there is any cloud cover above the stratus layer. However, when the stratus forms in an air mass which has a high moisture content to a considerable distance above the surface, the stratus will thicken as the mixing extends higher. The thicker stratus will retard solar heating, and so will contribute to its own power to persist. If the air is unstable, there will sometimes be a tendency for discrete cumuli to form, with a breaking up of the stratus layer. The layer of stratus will also be sometimes broken up by a very strong wind. Should the stratus layer be expected to thicken considerably, the possibility of precipitation in the form of drizzle must be considered.

Stratus formed late in the day may generally be expected to thicken at night, as a result of radiational cooling of the cloud's upper surface. Radiational effects sometimes give rise to a certain amount of instability. The cloud is cooled at its upper surface by outgoing radiation from itself, and is at the same time warmed from below by absorbing outgoing long wave radiation from the earth's surface. Some of this long wave radiation from the earth will be re-radiated by the cloud back to the earth.

Fractostratus forms in much the same ways as stratus, but generally under conditions such that the moisture content of the air is not sufficient to produce a complete layer of stratus, or the wind is too strong to permit the fractostratus elements to merge into a solid stratus layer. Should conditions be right for formation of stratus, fractostratus will generally form first, the elements later merging to form stratus. Lifting fog, when the lifting agent is solar heating, very often forms fractostratus rather than stratus. Such fractostratus is usually rather short lived. In almost all cases where large quantities of radiation fog are present or expected, formation of fractostratus when the fog lifts may successfully be forecast.

Fractocumulus may be expected to form when thermal or orographic updrafts are sufficient to raise the air to its convective condensation level. If the air contains only a small amount of water vapour, or the updrafts do not extend beyond a low level, the fractocumulus may be expected to remain such, otherwise it will continue to develop, forming cumulus.

The clouds of vertical development are most useful indicators of moisture content and stability conditions. Cumulus humilis indicates the existence of instability in the lower layers of the atmosphere and of updrafts which are thermal or orographic in origin. The updrafts must extend above the convective condensation level, which will be at a height which depends upon the moisture content of the air. When cumulus continues to develop, producing cumulus congestus, the forecaster knows that there must be a fairly deep layer of conditionally unstable air, whether the soundings show it or not. The point at which the upward development of the cloud is arrested provides an indication of the height of the base of the layer of absolutely stable air above. On the other hand, the development of the cloud may be arrested by the insufficiency of the supply of water vapour. When cumulus congestus is reported, it may be expected to be reasonably prevalent throughout the air mass, unless the stations reporting it are near hills or other such features conducive to greater development of cumuliform cloud than normal. Development of cumulus congestus to form cumulonimbus indicates that the layer of conditionally unstable air is very deep and that the air contains a very large amount of water vapour.

Alterations in the forms and sizes of clouds of vertical development provide useful indications of changes in the stability of the air. Such alterations, either observed locally or reported from other stations, merit careful atention from the forecaster.

Forecasting clouds of vertical development requires a careful study of stability conditions and of the water vapour content of the air. Such information may be obtained from upper air observations and from the history of cloud conditions in the air mass concerned.

If it is considered that the air is unstable throughout a rather shallow layer only, or contains only a small amount of water vapour, cumulus humilis may be forecast. The height of the base may be determined from a tephigram, but only very approximately. It is also possible, if the surface temperature and dew point are forecast for the time at which formation of cumulus is expected, to compute from these values the height at which the cumulus should form. This may be done by using the formula of Hann-Suring:

$$Z = 220 \ (T \text{ - } T_d) \ \text{feet,}$$

where Z is the height of the cloud base, T is the surface temperature, and T_d the surface dew point in degrees Fahrenheit. While this formula is only an approximation, the results obtained are reasonably satisfactory.

If there is a fairly deep layer of conditionally unstable air, and the air contains sufficient water vapour, cumulus congestus may be expected. Condensation may occur while the lapse rate is close to the dry adiabatic. Great quantities of energy are released, and the cloud builds upward rapidly. When it penetrates absolutely stable air above the layer of conditionally unstable air, energy is absorbed and the upward development is arrested. A study of the tephigrams from the nearest soundings and also a study of the cloud history of the air mass will help in deciding at what height the development will cease.

If the forecaster has reason to believe that the layer of conditionally unstable air is very deep and that its moisture content is high, he may expect that the cumulus congestus will continue to develop and form cumulonimbus. Advection aloft of unstable air containing a large amount of water vapour may cause development of cumulonimbus clouds. Such advection may occur in certain warm front situations and also in cases of convergence. Cumulonimbus may form in unstable air with much water vapour aloft while the air in the lower levels is still dry and stable.

In forecasting clouds of vertical development, very great care must be exercised to make allowance for changes, resulting from air mass modifications, in the stability and the moisture content of the air. Heating from below serves to steepen the lapse rate and so to cause instability, or to increase instability already present. This effect is very well illustrated in winter by the modification of fresh Polar Continental air flowing over the Great Lakes. The air is heated from below by the relatively warm water, the lapse rate is steepened, and instability results. Water vapour is also added from the lake surface by evaporation. In the lee of the lakes cumulus congestus and cumulonimbus develop, with snow showers. In summer, on the other hand, air passing over large bodies of water is cooled from below, its lapse rate being reduced and an increase in stability resulting. It is evident, then, that when considering the cloud history of the air mass the forecaster must take into account the modifications which may be reasonably expected to occur.

Topographical features are of the greatest importance in forecasting of cumuliform clouds, particularly the more active forms. If there are in the forecast district hills of appreciable size, these orographic effects are very considerable. In this connection, the strength of the winds to be expected is an important factor, since this very largely determines the extent of the updrafts the hills will cause. The degree of stability of the air is also an exceedingly important factor in determining the height to which the updrafts may be expected to extend. These factors must, of course, be considered in conjunction with a careful study of the moisture content of the air.

Rivers have very interesting effects on the paths of showers and thunderstorms of the air mass type. Even quite small rivers affect considerably the paths followed by air mass thunderstorms of small or moderate size. The downdraft over the river does not appear to feed the updraft of the storm satisfactorily, while over the land beside the river there is usually an adequate updraft. This effect has been frequently noted at the Ottawa airport, which is situated near the Rideau River. Such storms and showers tend to follow this relatively small river and pass by the edge of the airport. An empirical knowledge of the district is of great assistance in such situations.

9. Clouds in Relation to Flying

CLOUDS exert a very considerable influence on all flying operations. Aviators make use of clouds as indicators of the weather to be expected. Flights are generally planned to avoid flying in cloud. Concealment in clouds constitutes a hazard in that it increases the risk of collision and also adds to navigational difficulties.

The cloud forms present indicate to an experienced observer what weather conditions are to be expected in the future. The aircraft operator should correlate his cloud observations with information contained in the forecast provided for the flight. By so doing, he will be able to check the accuracy of the forecast and to make such revisions as appear necessary in the light of direct observational data. Even if the aviator has not had an opportunity to obtain a forecast, he may obtain from a careful and intelligent study of the cloud forms he sees a very fair idea of the weather processes which are occurring or will be likely to occur along or near his line of flight.

The forms in which the clouds appear provide an indication of the degree of turbulence in the atmosphere. In general, heap or cumuliform clouds indicate rough air, and stratiform or layer clouds indicate smooth air. Clouds possessing the characteristics of both forms, as for example layers of stratocumulus or altocumulus, indicate slight degrees of turbulence. The degree of turbulence, in addition to being significant if flight in or through the cloud is contemplated, will give the pilot an indication of what type of precipitation, if any, will occur. From the more turbulent clouds larger drops may be expected.

Clouds provide a useful indication of humidity conditions in the atmosphere. The aviator will know, for example, that just below the bases of cumulus clouds the relative humidity of the air is very high. This knowledge will assist him in anticipating carburettor icing conditions. When fairly large cumulus clouds dissipate at their tops, instead of growing upward to form large towers, the air at cloud top level is known to contain only a small amount of water vapour. Lenticular cloud forms indicate the presence of a layer of air of high relative humidity at the height at which these clouds are observed.

While the implications of the more spectacular clouds, such as cumulonimbus, are quite easily understood by most pilots, the implications of frontal clouds are less likely to be realized. A study of the high and middle cloud forms, along and near the line of flight, frequently furnishes useful indications of the presence and character of overrunning air.

Flight directly towards an active warm front from the cold air side will encounter the characteristic array of warm front clouds, and these should indicate to the pilot what type of cloud and weather is ahead of him. The succession of cirrus, gradually thickening to cirrostratus and then to altostratus, indicates that precipitation is not far ahead. Should this occur at an earlier or later stage on the route than was forecast, the pilot will have to adjust the entire forecast to conform to the evident change of speed of the front. The aviator should expect that shortly after the altostratus becomes thick he will encounter virga and then rain or snow. In many instances, he will shortly be obliged to descend below low cloud or to fly between layers of cloud. As the transition to nimbostratus occurs, with a lowering cloud base, the aircraft will be forced to descend to a lower altitude or to enter the cloud.

Flight on top of warm front clouds will also be more or less characteristic, although the changes in the tops will not, in general, be so noticeable as those below. If warm front thunderstorms are present, their heads will show clearly, and avoiding the cumulonimbus should then be relatively simple. Should the flight on top be from east to west, the front's position will be quite clearly shown by a sudden ending of the cloud, or by a very sharp decrease in its vertical extent. From very thick nimbostratus the cloud below will change to warm sector stratus or to no cloud at all. If advection fog occurs in the warm sector, it will appear from above the same as stratus cloud. The showers and thunderstorms observed in a warm sector where the air is unstable and has a high water vapour content are distinctly different from frontal clouds, since they do not appear in a solid array.

Flight in the cold air away from a warm front will show the characteristic array of clouds in reverse order. The lowest clouds disappear from below the aircraft, or above if it is flying at a low level in sight of the ground. The cloud base becomes gradually higher, and the cloud changes from nimbostratus to altostratus. The precipitation becomes intermittent, is reduced to virga only, and finally ceases. The altostratus becomes thinner and changes to cirrostratus thinning to cirrus, which gradually disappears.

Flight towards a warm front from the warm sector will show from above a great mass of cloud, with a fairly uniform top.

Convergence occurring along a trough of low pressure, no fronts being present, results in a sequence of clouds and weather very closely resembling, both from below and from above, that occurring at a warm front.

Approaching an active cold front from the warm sector, an aviator sees a large mass of cloud of considerable vertical extent. If the warm sector air is unstable and contains a large amount of water vapour, the clouds will be cumulonimbus. Such an array of clouds is generally a very severe hazard to flight. In advance of the vertical development clouds, there will generally be a well-organized layer of altocumulus. The appearance of a cold front from the cold sector will not be very different, except for the absence of the altocumulus in some cases. Again, changes in the speed of the front will be shown by occurrence of clouds at different points on the route from those expected.

An occlusion presents to the pilot's view a cloud system of its own. In an east to west flight through such a system he will observe cirrus increasing and thickening to form cirrostratus, which in turn thickens to altostratus. Precipitation may fall from the altostratus, frequently forming low cloud by evaporation from raindrops and re-condensation. Nimbostratus may next be encountered, with increased intensity of precipitation, and a lowering cloud base. In many cases, cumulonimbus is present, concealed in the nimbostratus. The cloud mass ends rather abruptly on the western side of the system. Here there are commonly considerable sheets of altocumulus clouds. Both types of occlusion have this general appearance from the air. The extent of the vertical development of the clouds depends very largely upon the moisture content of the air, and also upon the speed of motion of the system. Flights from west to east will encounter the same clouds in the reverse order.

Flight conditions through the northern part of a depression depend upon the distance from the centre and the degree of activity of the low. At a long distance from the centre, the aviator will observe considerable amounts of cirrus, and if slightly nearer, cirrostratus. Still closer, altostratus or nimbostratus and frequently precipitation will be observed, with, in the case of precipitation, extensive lowering of the cloud base.

On long-range flights, such as transoceanic journeys, entire depressions, with their frontal cloud systems, may not uncommonly be observed. They may be "topped" and viewed from above, or they may be to one side of the flight path, in which case a lateral view is seen.

Flight plans on the airways must be made to conform with existing and expected cloud conditions. The pilot will generally plan his flight to avoid, so far as possible, cloud flying. He will endeavour to fly within sight of the ground, on top of the clouds, or between layers of cloud. For many private pilots, navigation by visual reference to the ground is the only choice, since they do not have suitable instruments for cloud flying. Navigation on top, between layers, or in cloud will usually require radio equipment. The presence of a cloud layer will often prevent a pilot from taking advantage of favourable winds at the cloud level.

There are several properties of clouds which affect flight in or above them. Continuity, turbulence, density, temperature, height of base, and thickness all play important parts.

The continuity of the cloud is a matter of considerable interest. The pilot will welcome breaks in the cloud in or above which he is flying. These breaks, if sufficiently numerous, may even make possible visual flight in or above the cloud layer. In any case, sizeable breaks assist navigation by permitting visual identification of "pin points." This will be particularly helpful if the navigation is being carried out by dead reckoning.

The degree of turbulence is important if cloud flying is being undertaken.

Rough air increases the difficulty of instrument flying. In airline operations, where passenger comfort is an important consideration, the degree of turbulence has additional significance.

In planning cloud flying, consideration must be given to the temperatures in the cloud. The presence of supercooled droplets of water in clouds is very common, and flight in such clouds results in formation of ice on the aircraft. The rate at which ice will form is determined very largely by the temperature of the droplets. In this connection, the degree of turbulence is important, since in the more turbulent clouds larger droplets, supercooled to lower temperatures, are present. In non-turbulent clouds the droplets are usually smaller, and the icing hazard not so great. The density of the cloud affects the rate of ice formation, for it determines how many droplets will strike the aircraft. In clouds composed of ice crystals only, the airframe icing hazard does not exist. Carburettor icing is a danger outside of clouds as well as in them, and cannot be considered a difficulty peculiar to cloud flying.

Density of a cloud will not be a very great concern for the pilot of an aircraft flying in or above the cloud, except that it will be a factor in determining the altitude at which visual contact with the ground will be established on letting down. Occasionally, when flying above very diffuse clouds, the pilot may remain in sight of the ground.

The height of the cloud base determines the maximum height at which visual flight may be maintained and the height at which visual flight will be re-established after flight in or above the cloud. When flight in or above cloud is planned the height of the cloud base relative to the terrain is a most important factor in determining the level at which the aircraft may be operated with safety. The height above the terrain is also important in connection with letting down below the cloud base. For visual flight the height of the cloud relative to hills will often be the factor which determines whether or not the flight can be carried out with safety.

Thickness of a cloud layer determines how much time must be spent in it when climbing or letting down through it. This time factor is particularly important when icing conditions prevail in the cloud. The effective thickness of a cloud from the point of view of time varies considerably with the rate of climb or descent, which in turn varies with different types of aircraft, loads carried, and the manner of flying the aircraft.

High clouds do not usually affect flying operations to any appreciable extent. They are composed mostly of ice crystals and so do not constitute an icing hazard. Some cirrocumulus clouds are composed of water droplets, but the cloud is not dense and the droplets are so small and so cold that they will not cause an appreciable risk of airframe icing. High clouds are much too high for flight in them to involve any danger of collision with hills. Aviators find high clouds very useful in watching the approach of overrunning warm air.

Altocumulus does not in general present the pilot with serious problems. It is composed of ice crystals or of water droplets, the latter being commonly super-cooled. However, these will often be too small and too cold to cause any serious airframe icing hazard. In flight below altocumulus, virga are sometimes en-countered, and on rare occasions, rain or snow. Altocumulus is not generally continuous, having many interstices between elements and many thin spots.

Altostratus, in its thinner forms, is composed of ice crystals and so does not constitute an icing hazard. The thicker forms of altostratus contain supercooled water and sometimes cause quite appreciable ice accretion on aircraft. The density of altostratus translucidus is small, so that even when it is composed of supercooled water the rate of ice formation is not great. Altostratus is generally continuous for long distances. Flight below altostratus commonly encounters precipitation. Altostratus precipitans is considerably denser than altostratus trans-lucidus, and the icing hazard is correspondingly greater.

Both middle cloud forms are in most cases well above the highest terrain. Layers of cloud consisting of mixtures of altocumulus and altostratus are con-tinuous and frequently produce precipitation. The icing properties of such mixtures will closely resemble those of the thicker forms of altostratus.

Overrunning is indicated by extensive layers of altostratus. The stability con-ditions of the atmosphere are sometimes indicated by the form of the middle clouds. For example, altocumulus castellatus provides a useful indication of a small degree of instability at the level at which it forms. Altostratus provides the aviator with a useful indication of approaching precipitation.

Low clouds frequently constitute serious hazards to flying. The most serious is the proximity of the cloud base to the ground. Some of the low forms envelop hills in such a way as to cause a very real danger of collision with the hills. Low clouds are generally denser than the middle and high forms and so when icing temperatures prevail the rate of ice accumulation is greater.

Development of altostratus into nimbostratus creates a hazard which may be serious to an aviator flying in the altostratus. The cloud base becomes much lower, and the let down at the destination is rendered much more hazardous. The danger of the let down is increased by the fact that there will often be very low stratus or fractostratus below the nimbostratus or merged with it, the low cloud being formed in the rain which usually falls from nimbostratus. If the pre-cipitation is in the form of snow, the precipitation itself is a serious hazard when letting down. Nimbostratus is frequently turbulent enough to make instrument flying or formation flying difficult. The difficulty of formation flying is further increased by the great density of this cloud. Flight below nimbostratus is virtually certain to encounter rain or snow in considerable quantities.

Stratus is often very low, so low that visual flight below it is impossible, except over very flat terrain or over the sea. It frequently envelops hills, thus making let downs most hazardous unless the aviator knows his location and altitude exactly. The cloud is generally very smooth. The most usual and satisfactory flight path in regions where there is much stratus is on top or between layers with a let down at the destination. Stratus is so dense that a view of the ground is obtained from only a very short distance above the base.

Fractostratus does not usually constitute so serious a hazard to flight as stratus. There is, however, the very considerable danger that the fractostratus elements will fuse together into a layer of stratus. It is sometimes possible to maintain visual flight from a height well above the top of a layer of fractostratus, on account of the lack of continuity of the cloud. Low fractostratus may conceal hills, but the lack of continuity reduces this hazard. The cloud is not usually very dense, but it will cause a rather dangerous situation in the vicinity of an airport, particularly on the approach.

Stratocumulus rarely constitutes a serious hazard. Its vertical extent is usually small enough that flight paths may be selected above it or below it. The interstices between elements of some forms assist navigation from above the cloud, or in it. Most stratocumulus clouds have a sufficiently high base to permit visual flight below them. Stratocumulus is not generally a precipitation cloud, but may produce enough snow to render flight hazardous. When heavy cumulus or cumulonimbus clouds are concealed in the stratocumulus, the danger of precipitation becomes much more serious.

In itself cumulus humilis affects flying very little. However, it serves as a most useful indicator of the presence of updrafts. To the airline pilot a layer of cumulus indicates that flight above the cloud tops will be more comfortable for his passengers. To the sailplane pilot cumulus clouds indicate the presence of updrafts which he may use to gain altitude.

Cumulus congestus indicates more violent activity and may be used as an indicator of probable shower conditions. Since this form of cumulus often penetrates above the freezing level, ice may form on aircraft in its upper parts.

Cumulonimbus is an exceedingly dangerous cloud to the aviator. The extremely strong vertical currents and consequent great turbulence result in very difficult and dangerous flight conditions. The terrific velocity gradients subject aircraft to stresses far greater than those for which they are designed, and major structural failure sometimes results. The turbulence frequently takes the aircraft out of the pilot's control. Recent investigations by the Thunderstorm Project in the United States appear, however, to indicate that, in many thunderstorms, the turbulence is somewhat less severe than had been supposed. Extensive damage from hail is sometimes suffered by aircraft flying in and below these clouds, or below the anvil of a cumulonimbus incus. The large number of supercooled water drops, often of considerable size, make the icing hazard serious. Damage to modern aircraft by lightning is generally negligible, as an all-metal aircraft with proper bonding will carry the discharge without sustaining harm. Serious damage to radio equipment or complete destruction of that equipment results when the discharge or a part of it passes through the radio. This may often be successfully avoided by grounding the antenna, and reeling it in, if it is trailing. Occasionally, lightning destroys the pitot head, with serious consequences to the functioning of the flight instruments. Personal injury to aircraft crews in thunderstorms is very rare. Danger of losing control of the aircraft is sometimes the result of temporary blindness induced by the brilliant flash of near-by lightning. This is usually offset by lighting the cockpit of the aircraft.

Air mass thunderstorms may usually be circumnavigated without difficulty. Frontal storms, however, often are more serious. The tops of cumulonimbus clouds are frequently much too high for flight over them to be successfully attempted, except in aircraft capable of operating at high altitudes. Flight through a thunderstorm is generally undertaken only as a last resort.

Flight in layer clouds is hazardous when cumulonimbus clouds are concealed in the layer cloud. This is particularly likely to happen in warm front thunderstorms. Air mass thunderstorms are also sometimes found hidden from view at their bases in stratocumulus. Large numbers of cumulus clouds may make detection of the presence of cumulonimbus difficult. From high levels these clouds may often be seen from very great distances, on account of their enormous height. A warning of the presence of thunderstorms in the vicinity is often provided by "crash static" on the radio.

10. Stratosphere Clouds

IN THE foregoing chapters consideration has been given to cloud types which are observed within the troposphere. Certain other cloud forms are observed on very rare occasions at much higher levels, well within the stratosphere. These formations, known as *mother-of-pearl* or *nacreous* clouds and *noctilucent* or *night luminous* clouds, occur at heights of 22 to 30 kilometres and 80 to 90 kilometres respectively. Because these clouds occur at heights at which there is generally supposed to be little or no water vapour, their composition is as yet undiscovered.

These very high cloud forms may be the means of ascertaining new facts about the upper atmosphere. There is really very little known about the region where the noctilucent clouds occur, and there are very few ways of obtaining information about it. The region where the mother-of-pearl clouds occur is not so remote, but is still very difficult to investigate. Further, the clouds themselves are striking and beautiful phenomena.

NOCTILUCENT CLOUDS

Vestine obtained some very fine photographs of noctilucent clouds in Alberta in 1933. He made a comprehensive study of other recorded occurrences and of the state of knowledge concerning them.

A method for determining the minimum possible altitude from a single observation has been developed by Vestine. From this he computes minimum altitudes for the clouds observed at Meanook, Alberta. The results of the computations are 51, 57, and 55.5 kilometres. Altitudes of noctilucent clouds have been determined by means of simultaneous photogrphs taken from two stations a known distance apart, against the same background of stars. There is very good agreement between results obtained by different observers. In the summer months of 1889-91 Jesse obtained heights of 79 to 90 kilometres, the weighted mean of the observations being 82.08 kilometres. In 1932 Störmer obtained a height of 81.8 kilometres on July 10 and 81.1 kilometres on July 24. The average of 37 heights from 7 pairs of plates obtained by Störmer was 81.1 kilometres.

Noctilucent clouds are very tenuous, having forms very like cirrus and cirrostratus. Rapid changes in form occur, details being unrecognizable in a few minutes. Variations are observed in the intensity of illumination and in the colours. These variations are observed on different occasions and also during the same observation. During a display there appears to be a tendency for minimum intensity at about midnight, with increasing intensity thereafter. Colours of yellow, green, and orange are observed, as well as white. Variations in the colours are seen to occur with variations in the position of the clouds relative to the horizon. At low angular attitudes they have a golden or reddish-orange colour. Higher above the horizon they become white or bluish-white, and still higher they take on a blue-grey colour. The spectrum resembles that of daylight, having no emission lines, and showing the Fraunhofer lines clearly. The absence of red in the noctilucent clouds at high angular altitudes indicates that the light reflected from them has not for the most part passed through the troposphere. Light passing through the relatively dense troposphere would have the shorter wavelengths scattered, and show a preponderance of red. The light from these clouds has been described by some observers as polarized, but an observation by von Helmholtz furnishes evidence to the contrary. The particles of which the clouds are composed apparently scatter the blue light and transmit most of the red.

Statistical examination of reported occurrences indicates a maximum about 12 days after the summer solstice. It has been suggested that the prolonged continuous solar radiation at this season may be a factor in causing the formation of the clouds, if there are any chemical reactions involved. Observations appear to be confined chiefly to high latitudes.

The speed of motion has been computed on some occasions, very high speeds being indicated. Jesse reports a speed as high as 177 metres per second, which is equivalent to 400 miles per hour.

It appears quite possible that observations of noctilucent clouds would be more numerous if it were not for some of the occurrences being erroneously reported as aurora. The number of occurrences of aurora reported at the times of year most favoured by noctilucent clouds lends support to this view. Observations of aurora reported at times when noctilucent clouds are reported elsewhere also appear to favour this theory.

The nature of the material or materials of which these clouds are composed is so far unknown. They are attributed by Jesse to "foreign matter" at a height of about 82 kilometres, the foreign matter being the product of condensation of gases hurled to great heights by volcanic eruptions. In 1926 Wegener and Jardetzky, and in 1933 Vegard have expressed the opinion that they are composed of ice crystals. This theory would require a lower temperature at the noctilucent cloud level than is at present supposed to prevail. Humphreys has computed a temperature of $-113°C.$ as necessary for formation of ice crystal clouds at this altitude. The most recent work on the upper atmosphere indicates that the temperature prevailing is about $27°C.$ Upper air temperatures are not at present known with a sufficient degree of certainty that any theory regarding the materials of noctilucent clouds may be accepted or rejected on the basis of temperature conditions alone. In view, however, of the temperatures believed to prevail at noctilucent cloud levels it appears much more likely that if the clouds are made of water at all it is in the form of liquid droplets.

Dust particles have also been suggested as the material of the noctilucent clouds. In view of the scattering of the blue light, a size of 2×10^{-5} centimetres has been computed for the particles. The D-line in the spectrum of the night sky indicates that it is probable that atoms of sodium occur at a height of 60 kilometres. This may be due to ocean salt particles at heights much greater than was formerly suspected. If so, the salt particles themselves might be the material of the clouds, or they might be nuclei of condensation or of sublimation.

If the noctilucent clouds are composed of any form of water, there must be water vapour at heights in the atmosphere much greater than those to which it is generally supposed to penetrate. If the water vapour once succeeds in diffusing upward to the levels of high temperatures at all, there seems to be little or no reason to suppose that it would fail to diffuse upward to the noctilucent cloud level. Such a penetration of water vapour to the upper atmosphere would require diffusion through about 15 kilometres of very cold air above the tropopause. Should the assumption of diffusion of water vapour through the tropopause be found untenable, it may perhaps be that the water vapour reaches higher levels in latitudes where the tropopause is higher, and remains at this higher level. This would still involve passage of the water vapour through the tropopause, but it is hard to believe that the tropopause is such a substantial barrier as to prohibit this latter process.

Dust from volcanic eruptions has been suggested as the material of noctilucent clouds. It seems, however, quite doubtful if this volcanic dust could remain suspended at such very high altitudes for sufficiently long periods of time. The relative times of occurrence of volcanic eruptions and of noctilucent clouds are such that the dust would have to remain suspended for periods of several years. Attempts were made to link a particularly large number of displays in the period 1885 to 1894 with the eruption of Krakatoa in 1883. There does not appear to be any basis for ascribing occurrences in 1899 and in the period 1908 to 1911 to volcanic dust. An increase in observations of these clouds in 1932-33 followed a considerable number of minor volcanic eruptions, but these eruptions were not necessarily responsible. Further, it seems rather less likely that the minor eruptions would force the dust to such great heights. An eruption in Alaska in 1912, comparable to that of Krakatoa, does not appear to have coincided with any appreciable display of noctilucent clouds.

The earth's atmosphere may be affected by finely divided cosmic material near the orbits of comets and meteor showers, and this may be responsible for noctilucent clouds. Such a meteoric origin was suggested by Malsch and Backhouse. At the time of the Great Siberian Meteor in 1908 brilliant displays of noctilucent clouds were observed. This tended to confirm the theory of meteoric origin. That cosmic material is present at the heights where these clouds occur is well established. While the occurrences have not necessarily followed volcanic eruptions, they have followed known cosmic phenomena. In the period from 1880 to 1887 a large number of near cosmic appearances occurred, and it is quite possible that the large number of occurrences of noctilucent clouds in the period 1885 to 1894 may have been caused by material left in the atmosphere by these cosmic phenomena. Cosmic phenomena appear to offer a more probable source of the material of these clouds than volcanic eruptions. The material would merely have to settle, and would not necessarily be required to remain suspended for such great periods of time. Further, it is not necessary to assume any explosive force sufficient to project the material to great heights.

The evidence of the spectrum of the night sky that sodium is present at heights of the order of 60 kilometres suggests another possible source of the material of these clouds. However, it would seem most difficult for sea-salt particles to reach such altitudes, especially in view of the smallness of the convective activity in the stratosphere. It seems much more likely that water vapour would diffuse upward than that particles of salt would be carried up by air motions. This carrying of salt appears even less likely by far than the projecting to great heights of dust in volcanic eruptions. Meteorites are known to contain sodium, the amount being 0.17 per cent. It is reasonable to assume that the dust remaining from meteors consumed by burning will also contain sodium. This appears to be a rather more attractive hypothesis to account for the D-line in the spectrum of the night sky.

MOTHER-OF-PEARL CLOUDS

An outline of investigations in connection with these clouds was published by A. Thomson in 1932. They are very beautiful but are infrequently observed.

Mother-of-pearl clouds occur at measured altitudes of 22 kilometres to 30 kilometres. In the latitude of Oslo, where almost all of the observations of these clouds have been made, the highest measured cirrus clouds were at 11.34 kilometres. The average height of the tropopause in that locality is 10 kilometres. Störmer, using his technique for measuring the height of the aurora, obtained altitudes for the mother-of-pearl clouds of 26 to 30 kilometres.

Mother-of-pearl clouds assume varied forms somewhat resembling altocumulus; they are sometimes lenticular, sometimes of a rhomboidal form, and occasionally ribbon-shaped. They were described by Störmer in 1927 as having "beautiful rainbow colours." After sunset they become more brilliant and take on a strong red colouration, which dies away leaving a weak blue-grey. On one occasion they were observed at Oslo to persist all day as large iridescent clouds, like "uniform wavy altocumulus" of grey and grey-blue colour, changing in places to faint red, violet, and green. After sunset, another type was observed, having sharp contours and glowing intensely with a series of colours from red and violet to yellow-green and blue.

The colours in mother-of-pearl clouds are pure spectral colours, and extend in bands into the centre of the cloud, not being confined to fringes. The colours are not so prominent in the daytime as at night. In daylight there is sometimes a thin bluish-white background, against which the colours shine faintly.

Mother-of-pearl clouds were first reported by Möhn in 1893. He reported seeing them at rare intervals from 1871 to 1892 in the vicinity of Oslo, Norway. The next published account was in 1927 when Störmer reported observing them late in 1926, also at Oslo. At this time he mentioned having previously seen them in the nineties. Again in 1929 Störmer reported seeing displays. On this occasion they were observed before sunrise and again late in the afternoon and after sunset. Another observer reported the presence of large iridescent clouds over Oslo all day.

There is only one reported observation of these clouds from a point outside Norway. This observation was made from a train in the south of Scotland by Professor Irvine Masson on January 18, 1932.

The speed of motion of mother-of-pearl clouds was measured on December 30, 1926, and found to be 75 metres per second, equivalent to 170 miles per hour. On January 13, 1929, the speed was too small to be measurable, being approximately 1 metre per second. There is also considerable variation in observed vertical velocities. On January 13, 1929, the clouds were observed to fall steadily 1.5 kilometres in 79 minutes, which is equivalent to 0.3 metres per second. On the same day, another series of height observations indicated that for 21 minutes the height was constant at 23.8 kilometres.

Mother-of-pearl clouds have been observed only at times just after the passage of a wave of low pressure. In such circumstances the sky over level terrain is usually covered with thick clouds which prohibit observations of clouds at high levels. However, at Oslo the sky is cleared in such situations by the action of the föhn winds in the lee of the mountains of the Dovrefjeld range. It is reasonable to assume that occurrences of these clouds will in many cases be unobservable, except in regions where the skies are cleared by föhn effects in the rear of a wave of low pressure. If this is correct, northern Alberta should be a favourable location for observation of mother-of-pearl clouds.

Mother-of-pearl clouds occur in the stratosphere, where it is generally considered that there is little or no convective activity. Study of the propagation of sound indicates that the temperature of the air increases very rapidly at the level of the mother-of-pearl clouds. Duckert and Gutenberg report values in winter of −48°C. for 25 kilometres and 27°C. for 30 kilometres.

At such relatively high temperatures air is certainly capable of containing sufficient water vapour to produce clouds. The problem in connection with water vapour reaching these levels is essentially the same as that already outlined in connection with the noctilucent clouds. There would seem, however, to be rather more chance of the water vapour reaching the mother-of-pearl cloud level than of its reaching the noctilucent cloud level, for it would have 50 kilometres less atmosphere to penetrate. However, the problem still remains how it diffuses upward through about 15 kilometres of very cold air.

Störmer observed a corona in a quite homogeneous mother-of-pearl cloud. The corona consisted of a bluish-yellow inner region with an outer red ring, having an inside diameter of 14° or 15°. If the outer radius, which was not measured, is assumed to be 18°, the diameter of the cloud particles may be computed to be not more than 2.5×10^{-4} cm. This is of the order of size of the ordinary cloud droplets. The presence of coronae and iridescent colours in mother-of-pearl clouds is very strong evidence in favour of the clouds being composed of water droplets.

11. Optical and Electrical Phenomena Associated with Clouds

OPTICAL PHENOMENA

CLOUDS are responsible for several optical phenomena. These phenomena are caused by refraction, reflection, and diffraction of light by the components of the clouds and are useful indicators of the nature of the cloud components. Certain interesting effects of perspective are also observed, and some optical illusions.

In ice crystal clouds, *halos* are observed about the sun or moon. They are produced by refraction of light in ice crystals having the forms of hexagonal prisms, and floating with their axes horizontal. A faint halo shows as a ring of white light, a brighter one shows colours with red on the inner edge of the ring. The most common halo is that of 22 degrees, while the halo of 46 degrees is seen very rarely, and is usually incomplete.

Coloured or white images of the sun or moon are known respectively as *parhelia* and *paraselenae*. When coloured, they show red on the side towards the luminary. They appear at the same elevation as the sun or moon and often on or just outside the halo of 22 degrees. Images of this type at the same elevation as the sun, and having an azimuth distance from the sun of more than 90 degrees, are known as *paranthelia*. Like halos, these phenomena are produced by refraction, but they require the presence of ice crystals floating with their axes vertical.

Diffraction phenomena having the form of coloured rings are observed in water droplet clouds. *Coronae* appear around the sun or moon, having a red ring on the side farther from the luminary. When droplet sizes are varied, with no dominant size, the diffraction patterns overlap and blend. In this case, the only colours observed are in the region close to the sun or moon. A more or less fuzzy *aureole* results, having a brownish-red colour towards the outside and bluish-white towards the inside. Observation of these phenomena is generally easier when the luminary is the moon, because the sun produces excessively strong illumination.

It has been satisfactorily established by Simpson (1912) that coronae are produced by water droplets and not by ice crystals.

Tinted patches are sometimes observed in cloud, the colours being delicate red or green, occasionally blue or yellow. These colours occur in bands which tend to follow the outlines of the clouds. Simpson regards these bands of colour as fragments of large and bright coronae. *Irisations* in a cloud are evidence that it is composed of water droplets.

The *glory* is a phenomenon which is well known to aviators. Flying above clouds composed of water droplets one often sees a coloured ring on the top of the cloud, with the shadow of the aircraft in the centre of the ring. This phenomenon is caused by scattering of the incident light by the cloud droplets, and is observed exclusively in water clouds. A glory is occasionally seen by an observer on a mountain top with clouds below and the sun behind. Looking from a hill top away from the sun into a fog one sees a similar phenomenon. In these cases, one's own shadow appears in the centre of the ring or bow. The glory is also known as a *Brocken bow, Brocken spectre,* or *mountain spectre.*

A column of light, or *sun pillar,* above or occasionally below the sun is sometimes observed. This is the result of refraction of light by ice crystals suspended in the air. These ice crystals actually constitute a cloud, which is often very diffuse.

When the sun is low, bright rays and dark patches are seen to emanate from its position. These *crepuscular rays* are the result of sunlight illuminating dust particles in the atmosphere. The pattern of light and dark rays is caused by broken clouds, or by irregular terrain.

Cumuli viewed from a distance frequently have a deceptive appearance of being lower than those nearer the observer, when all are at the same height. Cloud bands, notably of high cloud, frequently have the appearance of being arched, and also of converging to a point, when actually they are horizontal. The former is an effect of *perspective,* the latter a species of *optical illusion.*

ELECTRICAL PHENOMENA

Electrical activity produced in clouds is responsible for some of the most spectacular displays associated with weather. Electrical discharges occurring between portions of the same cloud, between cloud and earth or between cloud and cloud take the form of enormous sparks, known as lightning. Large quantities of electrical energy are involved, and destructive results sometimes follow when cloud to earth discharges take place.

Streamers of light sometimes occur on aircraft and on other objects in the vicinity of thunderstorms. These produce spectacular effects on aircraft, forming rings around propellers and playing over the wings and other portions of the aircraft. This phenomenon is observed also on such objects as the masts of ships. *Brush discharge* or *St. Elmo's Fire* was observed by the ancients.

Observations with alti-electrographs indicate that the distribution of electric charges in a cumulonimbus cloud follows a fairly definite pattern. Above the 0°C. isotherm are found predominantly negative charges throughout a large portion of the cloud. In the upper regions positive charges are found to prevail. A positively charged region is found at the base of the cloud, located in that portion of the base from which the heaviest rain is falling.

To explain these charges numerous theories have been evolved. Of these only a few have survived. In 1885 Elster and Geitel published an "influence" theory. Falling raindrops may be expected to become charged by the earth's electrical field, their upper portions acquiring a negative charge and their lower halves a positive charge. When these drops fall, the air streams by them, carrying with it the cloud droplets. It is assumed that these cloud droplets make contact with the lower portion of the drop and then slide over its surface, remaining in contact, and leaving the drop's upper portion. The assumption is that they will leave with a negative charge acquired from their last contact with the drop. As a result the cloud droplets become negatively charged and the raindrops are left with a predominantly positive charge. In 1912 Elster and Geitel very drastically amended their theory. The revised theory assumes that the cloud particles do not slide along the surface of the drop at all, but instead rebound from it. It is assumed that during the contact with the lower surface the droplets acquire a positive charge, and the raindrops are left with a negative charge. It was pointed out by Simpson that electrical contact would not be made in such circumstances, on account of the interposition of the same film of air which acts to prevent fusion of drops on collision.

In 1905 Gerdien published the "Wilson-Gerdien condensation theory." This theory requires the existence of two extremely improbable conditions. Relative humidities of the order of 400 per cent are required, and also the absence of condensation nuclei. In such circumstances, condensation would occur on the negative ions. Even if such extremely improbable conditions were to exist, it does not appear possible that charges of anything like great enough magnitude could be produced in this way.

Simpson's "breaking drop theory" offers a most useful explanation of the charges found in the lower portions of thunder clouds. It is known that when a drop of water is broken by falling on a solid object a positive charge is acquired by the droplets resulting from the rupture, and the air becomes negatively charged. Simpson demonstrated that impact against a column of rising air will cause this separation of charges also. In the strong updrafts of a thundercloud this process is repeatedly occurring, the ruptures and recombinations resulting in the building up of large charges. This will explain the positive charge near the base of a cumulonimbus in the region of heaviest rain. The charges so caused are produced without the previous existence of an electrical field, such as is required for the operation of the "influence" processes.

In the upper portions of the cumulonimbus cloud a different process operates to produce charges. In the Antarctic, Simpson observed the development of large charges in snowstorms. This is the result of impacts of the ice crystals on each other. The ice becomes negatively charged, and the air positively charged. This same process occurs in the upper portions of cumulonimbus clouds. The negatively charged ice crystals will tend to sink more rapidly than the cloud droplets. As a result, the negative charges of the middle portions of the thunderstorms occur. The positive charge will tend to remain in the upper portions of the cloud, becoming attached to the cloud droplets. This process, like the breaking drop process, requires no previous electrical field.

In 1929 an "influence" theory was published by C. T. R. Wilson. This theory states that when raindrops are falling in the cloud with greater velocity than the positive ions, a separation of charges will occur. The positive ions in such cases will be unable to overtake the drops and so cannot neutralize the negative charges on their upper halves. The positive ions overtaken by the falling drops will be repelled by the positive charges on the lower portions of the drops. Negative ions will be attracted to the lower positively charged portions of the drops, and will tend to neutralize the positive charges, leaving the drops negatively charged. This method of building up charges requires the previous existence of an electrical field.

The breaking drop theory and the impact of ice crystals may be considered together to explain quite satisfactorily the electrification of thunderstorms. Much work remains to be done in connection with the electrification of clouds and precipitation. The charges on steady rain, for example, are a difficult unsolved problem.

Cloud Families and Forms. Their Heights (in feet).	Form	Colour	
HIGH CLOUDS Cirrus. 20,000 to 40,000 except in polar regions and during very cold weather.	Filaments, sometimes hooked; balls of snow with tails resembling commas.	Pure white. Appears grey when sun is low.	Diffus blur t or m shado of cir nitro-
Cirrostratus. 20,000 to 40,000.	Continuous uniform layer, with no relief; or closely interlaced fibrous streaks.	Pure white. Appears grey when sun is low.	Diffus outlin Thick conce
Cirrocumulus. 20,000 to 40,000.	Globules or ripples.	Pure white. Occasional coronae or irisations.	Diffus
MIDDLE CLOUDS Altocumulus. Summer, 8,000 to 20,000. Winter, 6,000 to 15,000.	Rounded elements or rolls. In lines or sheets.	White or grey.	Fairly exhib consi
Altostratus. Summer, 8,000 to 20,000. Winter, 6,000 to 15,000.	Uniform sheet, with very little or no relief. Precipitating form has diffuse base. Fibrous structure in thinner forms and in thin spots of thicker forms.	Grey, varying from very light to dark.	Varia um.
LOW CLOUDS Stratus. Near ground to 6,000. Frequently covers hilltops.	Uniform layer, often with ragged base. No regular relief features.	Light to medium grey.	Dense
Nimbostratus. Near surface to 6,000.	Uniform layer. No regular relief features.	Medium grey.	Very

	Thickness	Components and Their Sizes	Formation Processes
de nor he sun ral no ie case nd to-	Thin. Few hundred to a thousand feet.	Ice crystals only, often in form of hexagonal prisms.	Overrunning, either frontal or in zone of convergence. Tonitro-cirrus formed by heads of cumulonimbus broken away.
: blur moon. ur or	Thin. Not so thin as cirrus. Few hundred to about 1500 ft. Blue sky sometimes shows through complete very thin layer.	Ice crystals only, often in form of hexagonal prisms.	Overrunning, either frontal or in zone of convergence.
	Thin. Few hundred feet.	Usually ice crystals. On rare occasions supercooled water droplets.	Degeneration of cirrostratus or direct formation in rather turbulent overrunning air.
general Varies	Fairly thin. 100 ft. to 3,000.	Water droplets, ice crystals or snowflakes.	Frontal-lifting. Convergence. Altocumulus cumulogenitus formed by spreading out of tops of cumulus, the bases of the cumulus being dissolved.
medi-	Very thin, about 100 feet to thick, 6,000 to 8,000. Thinner forms, sun shows as through ground glass. Thicker forms, sun completely hidden.	Ice crystals in thin forms and in upper portions of thicker forms. Water droplets in thicker forms.	Frontal lifting. Lifting resulting from convergence.
	Thin to moderate. 50 ft. to 3,000 ft.	Water droplets. Favoured size of droplets 4 microns. Occasionally ice crystals in very cold weather. When composed of ice crystals, the cloud is usually very low cirrostratus and not stratus.	Orographic lifting. Formed in rain below precipitating altostratus and nimbostratus. Mixing; in this case it is usually the result of the merging of fractostratus elements. Lifting fog.
	Thick to very thick. 4,000 to 20,000.	Top portions ice crystals. Lower portions mixtures of water droplets and ice crystals. Favoured droplet size 10 microns.	Large scale frontal or convergence lifting, or orographic lifting of considerable extent.

Precipitation	Distinguishing Features	Height Determination	
None.	Fibrous elements do not form a continuous layer as in cirrostratus. When head of cumulonimbus becomes detached it is tonitro-cirrus, and for a time retains the characteristic "anvil" shape.	Estimation is difficult. May be determined by means of aircraft or balloons.	At co M au
None.	Continuous or very nearly continuous as opposed to non-continuity of cirrus.	Estimation is difficult. May be determined by means of aircraft or balloons.	Oc fai ill dif
None.	Convention requires that it be associated with ci or cs. This eliminates delicate edges of altocumulus sheets, which may easily be confused with cc.	Estimation is difficult. May be measured by means of aircraft or by balloons. Better adapted to estimation than other cirriform clouds.	Co de
Mostly none. Rarely light rain or snow. Virga fairly common.	Distinguished from sc by size of elements; the smallest regularly arranged elements of altocumulus less than 10 solar diameters (judged by 3 fingers at arm's length).	May be readily estimated. Aircraft reports. Balloons, pilot, and ceiling. Radiosondes. Projectors.	W Se fo co sa
Light rain. Light to moderate snow.	Grey colour distinguishes it from cs. Appearance of sun and moon also is different; their outlines are blurred.	Not suitable for estimation. Thinner forms or forms having ac patches may sometimes be estimated. May be measured by means of aircraft or by balloons, latter method not satisfactory in precipitation.	
Drizzle or very light fine rain or snow.	Absence of regular structure.	Measurement by balloon, projector, or aircraft. Estimation generally impossible, except with aid of hill or object of known height.	
Continuous rain or snow; of intensity varying from light to heavy.	No regular structure. Distinguished from stratus by precipitation which is in form of real rain or snow, not drizzle.	Measurement; estimation usually impossible. Balloon unsatisfactory on account of precipitation. Projector or aircraft measurement best. Comparison with high ground or object of known height.	Ap na

atures	Turbulence	Optical Phenomena	Electrical Phenomena
occasionally aint aurora. mination of ference.	Effectively none.	Occasional halo— not common.	None.
nfused with Motion and aurora show	None.	Halo, parhelia, paraselenae, sun pillar.	None.
uently with ac sheets.	Very slight. More than other cirri- form clouds.	Occasional coronae.	None.
of forms. of different erent levels rved at the	Slight.	Irisations and coronae common.	None.
	None or very slight.	Rare halo, in thinner forms.	None.
	None.	Glory when viewed from above, occa- sional corona or a u r e o l e w h e n viewed from below.	None.
hough illumi- de.	Considerable.	None.	None. Cumulonimbus with accompanying thunderstorm activity is sometimes concealed in ns.

Cloud Families and Forms. Their Heights (in feet)	Form	Colour	
LOW CLOUDS Stratocumulus. 1,000 to 6,000. Much higher in mountainous regions.	Rounded elements or rolls, frequently with wave-like structure, showing crests and troughs. Structure generally regular.	Grey.	Dense
Fractostratus. Near ground to 6,000.	Very irregular.	Grey.	Fairly
Fractocumulus. 1,000 to 6,000.	Lumps, and slightly rounded elongated wisps.	White to light grey.	Mode
CLOUDS OF VERTICAL DEVELOPMENT. Cumulus humilis. Bases 500 to 10,000. Tops 1,000 to 15,000. Often much higher in mountainous regions.	Flat base, rounded top.	Viewed from above white, often very brilliant. Viewed from below, grey. Darker in central portions. Appear much darker when there is a layer of high or middle cloud above.	Dense
Cumulus congestus. Bases, 1,000 to 6,000. Tops, 6,000 to 15,000. Often much higher in mountainous regions.	Flat base, turbulent top—frequently with towers.	White from above. Grey from below, sometimes very dark grey.	Dense
Cumulonimbus. Bases 1,000 to 6,000. Tops 10,000 to 40,000. Higher bases in mountainous regions, higher tops in tropical regions.	Turbulent, ragged base, cirriform top. Cumulonimbus calvus—no anvil. Cumulonimbus capillatus — cirriform cap. Cumulonimbus incus — anvil-shaped cirriform top.	White from above and from high levels beside cloud. Very dark grey to almost black from below. Lateral view, dark with white top.	Very tions, diffuse riform

	Thickness	Components and Their Sizes	Formation Processes
	Thin to medium. Few hundred feet to 2,000 to 3,000.	Water droplets. Favoured size 8 microns. Favoured size for roll form 7 microns.	Vertical mixing—in this case sc may be the result of a change from st or fs. Merging of cumuli to form sc cumulogenitus. Spreading and merging of dissolving cumuli late in the afternoon to form stratocumulus vesperalis. Orographic lifting. Addition of water vapour in rain, combined usually with vertical mixing.
	Thin. 50 ft. to 1,000 ft.	Water droplets. Occasionally ice crystals in very cold weather.	Vertical mixing. Addition of water vapour in rain—with or without mixing. Orographic lifting. Lifting of fog by turbulence.
	Thin.	Water droplets.	Thermal lifting. Orographic lifting.
	Thin to medium. 500 to 5,000 ft.	Water droplets. Favoured size 6 microns.	Thermal or orographic lifting. Develops from fractocumulus.
	Thick to very thick. 5,000 to 15,000 ft.	Water droplets.	Thermal or orographic lifting. Frontal lifting. Develops from cumulus humilis.
er por- l fairly or cir-	Very thick. 10,000 to 40,000 ft. Thicker in tropical regions.	Water drops and droplets in lower portions. Mixture of water and ice in higher portions. Top composed of ice crystals only. Hail in various portions of cloud on occasion.	Thermal, orographic, or frontal lifting. Requires unstable air with large water vapour content. When cumulus congestus extends to well above the freezing level, cb is formed.

Precipitation	Distinguishing Feature	Height Determina
Usually none. Occasional very light rain or light snow. When precipitation of greater intensity appears to fall from sc, it is falling from cb concealed in the sc.	Distinguished from ac by criterion that smallest regularly arranged elements are greater than 10 solar diameters in size (judged by width of 3 fingers at arm's length).	Very suitable for Measurement by project or aircraft. Comparison ground.
None.	Ragged form.	Measurement by balloon or projectors. Irregular be quite well adapted to of height. Comparison ground.
None.	Distinguished from fractostratus by more rounded form.	Estimation. Aircraft o Too much scattered fo Estimation from surface ture and dew point.
None.	No "towers." Does not have "boiling" top like cumulus congestus.	Estimation from surface ture and dew point. Vis tion. Aircraft observat jectors. Not generally a balloon height determi account of broken or nature.
Light showers of rain or snow from thicker forms.	Distinguished from cumulus humilis by heavy active tops or "towers," and from cb by absence of ice crystal structure of the top.	Estimation from temper dew point at the surfa estimation. Aircraft o Projectors. Not adapted determinations.
Moderate to heavy showers of rain or snow. Only cloud from which true hail falls.	Cirriform head identifies cumulonimbus. Occurrence of moderate or heavy showers indicates presence of cumulonimbus, even when head is concealed.	Estimation. Very danger tempt use of aircraft fo ment. Balloon unsatisfact count of violent vertica Heavy precipitation will vent determination.

Special Features	Turbulence	Optical Phenomena	Electrical Phenomena
	Slight, to moderate.	Coronae in thinner forms. Shows mostly around moon, as sun is too bright in many cases to show it. Crepuscular rays when sun shines through breaks.	None.
	Slight.	Parhelia in ice crystal forms. Occasional coronae in water droplet forms.	None.
	Slight to moderate.	Generally none.	None.
	Moderate.	Crepuscular rays under suitable lighting conditions—sun behind cloud and usually low.	None.
Sometimes causes formation of pileus.	Moderate to severe.	Crepuscular rays under suitable lighting conditions.	None.
Cirriform head, violent precipitation, and violent turbulence.	Very severe.	Generally none.	Lightning. Brush discharge (St. Elmo's Fire).

Cloud Families and Forms. Their Heights (in feet).	Form	Colour	

SOME SPECIAL FORMS
Lenticular
    Moazagotl. Approx.
    20,000.
    Ac lenticularis
    6,000 to 20,000
    Sc lenticularis
    1,000 to 6,000
    "Contessa del Vento".
    Pileus 6,000 to 15,000.
	Lenticular. Thin edges. Appear plate-like from below.
	
	"Contessa del Vento" is heavy form. Veil above and on tops of cumulus.
		Generally white or light grey. Heavier forms darker grey.
		
		White or light grey.
			Ge
			mo
			
			Sl

Banner. Varies with height of hill causing it.	"Banner" in lee of mountain top.	White to light grey.	M

Castellated
    Ac castellatus
    6,000 to 15,000.
 	
    Sc castellatus
    1,000 to 6,000.
	Narrow columns on common base—resembling battlements of a castle.	Acc white to light grey. Scc light grey to dark grey.	Ae
			de
			
			Sc

Mammilated
    Sc mammatus
    Cb mammatus
    1,000 to 20,000.
	Heavy "pouches" or "festoons" of cloud hanging below main cloud base.	Grey, usually dark.	De
			de

STRATOSPHERE CLOUDS
    Mother of Pearl or
        Nacreous
    22 to 30 km. 68,000 to
    92,000 ft.
	Like altocumulus "lenticular rhomboidal or ribbon-like forms." Also sometimes described as resembling uniform wavy altocumulus.	Bands of pure spectral colours. In daylight "thin bluish white background" against which colours shine faintly. Glow intensely after sunset, with series of colours from red and violet to yellow, green, and blue.	Sl

Noctilucent
    80 to 90 km. 240,000 to
    270,000 ft.
	Similar to cirrus and cirrostratus. Very tenuous. Rapid changes in form. Variations in intensity of illumination.	Yellow, green, and orange, as well as white. Variations in colours. Spectrum similar to daylight, shows Fraunhofer lines.	V

Thickness	*Components and Their Sizes*	*Formation Processes*
Slight.	Droplets of water or ice crystals, depending upon temperature at cloud level.	Indirect lifting. Layer of damp air lifted by raising of air below.
Thin—few hundred to about 1,000 to 2,000 ft.		Lifting caused by obstacles in the flow of air; mountains or "convective elements."
Few hundred to about 2,000 ft.		
Thick.	Water droplets.	
Thin, often less than 100 ft.	Water droplets.	Indirect lifting of moist air by developing cumulus.
Fairly thin, less than 100 ft. to about 1,000 ft.	Water droplets.	Remains of cloud formed on windward side, or cloud formed on lee side of hill from just sufficient lift.
Acc, few hundred to about 1,500 ft.	Water droplets or ice crystals.	Formed when updrafts are present whose cross-sections are small compared to their vertical extent.
Scc, 1,000 to 2,000 ft.	Water droplets.	
Mammilated surface usually occurs on base of thick cloud or underside of anvil of cb incus mammatus.	Water droplets.	Forms under conditions of great instability—cold air above warm air. Theories: Return of "past equilibrium" air which has risen. Waves on surface of discontinuity on lower side of anvil of cb.
Fairly thin, similar to altocumulus.	Nature unknown. Iridescence leads to belief that components are water droplets. Too warm for ice crystals at levels where these clouds occur.	Unknown. Appear to occur on rear of low.
Thin.	Nature unknown, likely either water droplets or cosmic dust. Appears to be too warm at these levels for ice crystals.	Unknown.

Precipitation	Distinguishing Features	Height Determination
None.	Lenticular form, heavy in case of "Contessa del Vento."	Measurement by aircraft mation.
None.	Delicate veil.	Measurement by aircraft. Estimation.
None.	"Banner" observed only on lee side of hills.	Measurement by aircraft. Estimation. Knowledge of height of hill cloud.
Acc, none. Scc, may develop into shower cumulus.	Characteristic flat base, with small "towers" or "turrets."	Measurement by aircraft. Estimation. Not usually extensive enoug adapted to use of balloons jectors.
Frequently part of shower or storm cloud.	Heavy "pouches" or "festoons."	Estimation or projector. Balloon very doubtful on ac vertical currents. Aircraft measurement hazar account of dangerous tu and hail.
None.	Luminous after sunset. Shine faintly in daytime.	Simultaneous photograph two stations, as for aurora determination.
None.	Luminous at night.	By simultaneous photograp two observing stations, usir technique as to determine h the aurora.

Special Features	Turbulence	Optical Phenomena	Electrical Phenomena
ᵢ—occur at intervals h appear to be wave s, in lee of mountains.	None or very slight.	Occasional irrisations in altocumulus lenticularis.	None.
just above or below ulus heads. "Cap" on above tops. "Scarf" t cloud "shoulders."	None or very slight.		None.
n of cloud and loca- relative to hill are acteristic.	Slight.		None.
base and turrets. quently is an indica- of instability at the l where it forms.	Slight to moderate.		None.
racteristic form. cates great instability.	Considerable.		Usually forms part of cumulonimbus, which produces lightning and brush discharge.
ninous at night. Shine tly in daytime. ibit pure spectral urs.	Unknown, likely slight.	Irisations.	None known.
ninosity at night. ours. at height.	Unknown, most likely very little or none.		None known.

GLOSSARY

Glossary

Adiabatic. An adiabatic process is one which takes place without any gain or loss of heat. The temperature, however, changes.

Advection. Transfer by motion in a more or less horizontal direction.

Aerosol. A suspension of a liquid in a gas; similar in many respects to a colloid made by suspending a liquid or a solid in a liquid.

Air mass. A relatively large portion of the atmosphere, having approximately homogeneous properties (in the horizontal) with respect to temperature and moisture content.

Alidade. An instrument for measuring vertical angles.

Amorphous. Lacking grain or other structural detail, not crystalline.

Atmosphere. The body of air which surrounds the earth, and is carried with it through space. It is a mixture of gases, in nearly constant proportions.

Aureole. A luminous region surrounding a source of light. The term may be applied to the apparently transparent space within a corona, or to a condition where a corona is not quite developed.

Aurora. A luminous display caused by electrical discharges at heights of the order of 100 kilometres. The light is frequently coloured.

Castellated. Having a form resembling the battlements of a castle.

Chinook. A warm dry wind on the eastern side of the Rocky Mountains.

Clinometer. An instrument for measuring vertical angles, different from the alidade in construction. This instrument is commonly held in the hand.

Coalescence. Joining by merging.

Colloid. A colloid is a suspension of one substance in another. The particles of the suspended substance (dispersed medium) are larger than molecules. This criterion distinguishes a colloid from a true solution. In chemical terminology, a colloid is usually understood to have suspended particles of the order of one micron in diameter.

Condensation. A change in state from vapour to liquid. When this change in state occurs, the latent heat of vapourization is released.

Convection. Convection is a transfer of heat by a transfer of the matter containing the heat. When a fluid (the term "fluid" includes gases, and mixtures of gases, such as air) is warmed, some of it expands and so becomes less dense than surrounding portions of the fluid. The result is that the warmed portion tends to rise.

Convergence. Said to occur when two currents of air flow together or, more generally, when horizontal flow causes the accumulation of air in any fixed volume in the atmosphere.

Corona. A coloured ring surrounding the sun or moon. The ring is red on the outside, and displays the spectral colours, changing to violet on the inside.

Crepuscular rays. Solar rays made visible by the sun shining on dust particles in the air, when most of the sunlight is cut off by clouds or mountains. These phenomena are observed when the sun is quite low, and also when it is below the horizon.

Dewpoint. The temperature at which air would be saturated with water vapour if cooled without adding or removing any water vapour.

Diffraction. The apparent bending of light around obstacles.

Evaporate. Change from liquid state to the vapour state. During this change the latent heat of vapourization is taken in by the substance whose state is changing.

Fog. Suspended water droplets or ice crystals in the air immediately above the earth's surface. A fog is effectively a cloud in contact with the ground.

Föhn. A warm wind blowing on the lea side of mountains. The air is warmed by compression in descending the mountains.

Fraunhöfer lines. Dark lines in the sun's spectrum, caused by absorption of certain wavelengths in the passage of the sun's rays through the sun's atmosphere.

Front. A narrow zone of rapid transition between air masses of different properties.

Glaciation. Freezing, changing to ice.

Halo. A coloured ring about a luminary, usually the sun or moon. It is caused by refraction of light. The spectral colours are so arranged that red is on the inner edge of the ring, and violet on the outer edge.

Haze. Very small water droplets or hygroscopic particles suspended in the atmosphere. The droplets are smaller than those of fog or cloud.

Hydrosol. A colloidal suspension in a liquid medium.

Hygroscopic. Having an affinity for water.

Interstice. An opening or clear space, as between parts of a cloud.

Inversion. A situation where the usual decrease of temperature with height is "inverted," and temperature increases with height.

Ion. A particle of matter carrying an electric charge.

Irisation, iridescence. The name given to patches of colour, such as are quite commonly observed in clouds composed of water proplets.

Isotherm. A line drawn on a chart through locations having the same temperature.

Kern. A nucleus.

Lamina. Layer.

Lapse rate. Rate of decrease of temperature with height. Rising and expanding air tends to cool at the dry adiabatic lapse rate until it becomes saturated, and then at the lesser saturated adiabatic lapse rate.

Latent heat. The amount of heat absorbed or lost during a change of state. For example, during the melting of ice, the "latent heat of fusion" is taken in by the melting ice.

Lenticular. Having the shape of a lens.

Mammilated. Having hanging pouches.

Micron. A commonly used term for 0.001 millimetre.

Mist. A term sometimes used for a light fog.

Moazagotl. A type of cloud formed on the lee side of mountains under föhn conditions. The name is a traditional German one, and is supposed to relate to the earliest observer of these clouds, Gottlieb Motz.

Nacreous. Pearly. This term is used for the type of cloud which is commonly called "mother of pearl."

Noctilucent. Night luminous. The clouds to which the term is applied are illuminated at night by the sun. This is possible because of their great height.

Nucleus (of condensation or sublimation). A "centre" or "object" on which water condenses or sublimes from the vapour state to the liquid or solid state.

Occlusion. When a cold front overtakes a warm front, and the warm air is lifted out of contact with the earth, the resulting front is called an occlusion.

Overrunning. Warm air is often forced to rise when it is displacing cold, denser air, or is being replaced by colder air. The warm air then is said to be "overrunning" the colder air.

Paranthelion. A luminous patch, or "mock sun," at the same angular elevation as the sun itself, but at an angular distance of 120 degrees from it.

Paraselenae. Luminous patches at the same angular elevation as the moon, and usually on a lunar halo. These are also known as "mock moons."

Parhelion. A luminous patch, or "mock sun" or "sun dog" at the same angular elevation as the sun and frequently on a 22 degree solar halo.

Phase. In connection with water, this term is used to mean the state of the water, that is, solid, liquid, or vapour.

Pileus. A small cloud, an aberrant form of altocumulus caused by rising air above a cumulus cloud.

Polarized light. This term is used to describe light which has by some means had the vibrations in all but one plane damped out.

Precipitation. If the term is used strictly, both condensation and sublimation to form cloud or fog, and also falling hydrometeors, are precipitation. The term is usually used to mean only falling rain or snow.

Radar. An electronic device for detecting objects by the reflections of electromagnetic waves. The term comes from the words, "radio detection and ranging."

Radiation (solar). The energy of the sun reaches the earth and other bodies by radiation.

Refraction. The change of direction of light when it passes from one medium to another of different optical density.

Relative humidity. The ratio of the actual water vapour content to the maximum possible at the prevailing temperature and pressure.

Ridge of high pressure. A relatively long narrow region where the pressure of the atmosphere is higher than in the surrounding regions.

Saturation. Air is said to be "saturated" when it contains the maximum possible amount of water vapour for its temperature and pressure.

Specific humidity. The weight of water vapour present per unit mass of moist air.

Spectrum. The term "spectrum" is used for the series of colours into which light is divided by refraction, as in a prism, at a grating or in a water drop.

Stability. In a meteorological sense, the term stability refers to the state of the air with respect to vertical motion. Stable air tends to settle back after a lifting force is removed; unstable air tends to continue to rise.

Stratosphere. The portion of the earth's atmosphere above the troposphere. The normal temperature decrease with height virtually ceases at the base of the stratosphere. The temperature, however, increases again at higher levels.

Striated. "Lined" or striped.

Sublimation. The change from vapour to solid or from solid to vapour without passing through the liquid state.

Subsidence. Sinking of air in the atmosphere is called subsidence. When air sinks in this way, it is warmed by adiabatic compression.

Supercooled. The term is used for a liquid which is cooled below its freezing point, but is still in the liquid state. The term "undercooled" is also used for this condition.

Supersaturation. Air containing more water vapour than it normally can hold at the prevailing temperature and pressure is "supersaturated." The term is used in chemistry to describe solutions which contain more than the saturation amount of solute.

Tephigram. A thermodynamic diagram used to study the stability and moisture conditions of the atmosphere. Temperature (T) is plotted against entropy (phi), hence the name.

Theodolite. An instrument for measuring horizontal and vertical angles.

Thermal. Pertaining to heat. The word is also used as a term for an updraft, consisting of rising warm air.

Troposphere. The lower portion of the earth's atmosphere, wherein temperature nominally decreases with height.

Updraft. Upward-moving air.

Vapour tension. The pressure exerted by the water vapour present in the air.

Virga. Precipitation falling from a cloud but evaporating without reaching the ground.

Water vapour. Water in the gaseous state.

BIBLIOGRAPHY

Bibliography

Abe, Count Mansanao. *Distribution and movement of cloud around Mt. Fuji, July 1932 to August 1933.* Toyko: Central Meteorological Observatory, 1937.

Abercromby, R. On the identity of cloud forms all over the world. *Q. J. Roy. Met. Soc.,* XIII (1887), 140–7.

Air Ministry Meteorological Office. *Cloud Forms.* London: H. M. Stationery Office, 1937.

—————— *Meteorological Glossary. Ibid.,* 1939.

Aitken, J. On Dust Fogs and Clouds. *Trans. Roy. Soc. Edin.,* XXX (1880).

Archenhold, F. S. Die leuchtenden Nachtwolken und bisher unveröffentlichten Messungen ihrer Geschwindigkeit. *Das Weltall,* XXVII (July, 1938).

Bergeron, T. On the Physics of Cloud and Precipitation. *Procès Verbaux de l'Association de Météorologie, International Union of Geodesy and Geophysics* (Lisbon, 1933).

Bigelow, F. H. *Report on the International Cloud Observations May 1, 1896 to July 1, 1897.* Washington, United States Weather Bureau Report, 1898-99, II.

Boylan, R. K. *Proc. Roy. Irish Acad.,* (A) XXXVII, 58–70.

Bricard, J. *La Météorologie* (mars-avril, 1939).

—————La Teneur des Nuages en Eau Condensée. *Ibid.* (janvier-juin, 1943), 57–69.

Brooks, C. F. *A Guide to Cloud Coding.*

Blue Hill Meteorological Observatory, Harvard University.

—————Coronas and Iridescent Clouds. *Monthly Weather Review,* LIII (Feb., 1925).

—————The Value of Cloud Observations. *Trans. Amer. Geophys. Union,* 1933.

Bureau, R. Altimétrie des Nuages par Impulsions Lumineuses. *La Météorologie* (juillet-septembre, 1946), 292–301.

Byers, H. R. *Synoptic and Aeronautical Meteorology.* New York: McGraw-Hill, 1937.

—————and R. R. Braham. Thunderstorm Structure and Circulation. *J. Met.,* V (June, 1948), 71–86.

—————and Richard D. Coons. The 'Bright Line' in Radar Cloud Echoes and its Probable Explanation. *Ibid.,* IV (June, 1947), 75–81.

Clarke, G. A. *Clouds.* New York: Dutton, 1920.

Clayden, A. W. *Cloud Studies.* London: Murray, 1st ed. 1905, 2nd ed. 1925.

Clayton, H. H. Cloud Observations. *Annals of the Astronomical Society of Harvard College,* XX (Cambridge, 1887), 50.

—————Discussion of the cloud observations made at the Blue Hill Observatory. *Ibid.,* XXX, part IV (Cambridge, 1896), 271-500.

—————Measurement of cloud heights, velocities, and directions. *Ibid.,* XLII, part II, Appendix C (Cambridge, 1900), 193-280.

————A study of clouds with data from kites. *Ibid.*, LXVIII, part III (Cambridge, 1911).

Conover, J. H. and S. H. Wollaston. Cloud systems of a winter cyclone. *J. Met.*, VI (Aug. 1949), 249-60.

Coons, Richard D., R. C. Gentry and Ross Gunn. *First Partial Report on the Artificial Production of Precipitation.* Washington, United States Weather Bureau Research Paper No. 30, Aug., 1948.

————Earl L. Jones, and Ross Gunn. *Second Partial Report on the Artificial Production of Precipitation, Cumuliform Clouds, Ohio, 1948. Ibid.* No. 31, Jan., 1949.

————*Artificial Production of Precipitation. Ibid.* No. 33, Sept., 1949.

————Fourth Partial Report on the Artificial Production of Precipitation: Cumulus Clouds, Gulf States, 1949. *Bull. Am. Met. Soc.*, XXX (Oct., 1949), 289–92.

Coste, J. H. and H. L. Wright. *Philosophical Magazine*, VII (1935, No. 20), 209–34.

Coulomb, J. and J. Loisel. *La Physique des Nuages.* Paris: Editions Albin Michel, 1940.

Craddock, J. M. The Development of Cumulus Cloud. *Q. J. Roy. Met. Soc.*, LXXV (April, 1949), 147-53.

Cunningham, R. M. A different explanation of the bright line. *J. Met.*, IV (Oct., 1947), 163.

Cwilong, B. M. Sublimation in a Wilson Chamber. *Nature*, CLV (Mar. 24, 1945), 361-2.

————Sublimation in Outdoor Air. *Ibid.*, CLX (Aug. 9, 1947), 198.

Deppermann, C. E. An Improved Mirror for Photography of the Whole Sky. *Bull. Am. Met. Soc.*, XXX (Oct., 1949), 282–5.

Douglas, C. K. M. Optical Phenomena and the Composition of Cloud. *Journal of the Scottish Meteorological Society*, XVIII (1918).

————The appearance of the Sun and Moon through a cloud. *Meteorological Magazine* (Jan., 1929).

Duckert, P. *Ergebnisse der Kosmischen Physik*, I (1931), 236-83.

Elster, J. and H. Geitel. *Annalen Physik und Chemie*, XXV (1885), 121–31.

————*Physikalische Zeitschrift*, XIV (1913), 1287–92.

Findeisen, W. Zur Frage der Regentropfenbildung in reinen Wasserwolken. *Met. Zeit.*, LVI (1939).

————Das Verdampfen der Wolken und Regentropfen. *Ibid.*

————Die kolloidmeteorologischen Vorgänge bei der Niederschlagsbildung. *Ibid.*, LV (1938), 121–33.

————Entstehen die Kondensationskerne an der Meeresoberfläche? *Ibid.*, LIV (1937), 377-9.

Flower, W. D. Cumulus cloud produced by a fire. *Q. J. Roy. Met. Soc.*, LXVI (July, 1940), 279.

Fraser, D. Production of Ice Crystal Clouds by Seeding. *Nature*, CLXIV (July 30, 1949), 179.

Gerdien, H. *Physikalische Zeitschrift*, VI, 647–66.

Green, H. L. Disperse Systems in Gases, Dust, Smokes and Fog. The Faraday Society, London, 1091–6.

Gutenberg. Der Aufbau der Atmosphäre. *Handbuch der Geophysik*, IX (1932).

Hann. J. von and R. Süring. *Lehrbuch der Meteorologie.* 5th ed., Leipzig: W. Keller, 1939. I, 345.

Helmholtz, R. *Met. Zeit.*, XXIV (1889), 186.

Heywood, G. S. P. Rain formation in

the tropics. *Q. J. Roy. Met. Soc.*, LXVI (Jan., 1940), 46.

Hewson, E. W. and R. W. Longley. *Meteorology Theoretical and Applied.* New York: Wiley, 1944.

Howard, Luke. *Essay on the Modifications of Clouds.* 3rd ed., London: John Churchill, 1865.

————On the Modifications of Clouds. *Philosophical Magazine* (1803).

Humphreys, W. J. *Physics of the Air.* 3rd ed., New York: McGraw-Hill, 1940.

————Nacreous and Noctilucent Clouds. *Monthly Weather Review,* LXI (1933), 228.

International Meteorological Committee. *International Atlas of Clouds and States of the Sky.* Paris, 1933.

Jacobs, Woodrow C. Preliminary Report on a Study of Atmospheric Chlorides. *Monthly Weather Review,* LXV (April, 1937), 147.

Jardetzsky, W. *Met. Zeit.,* LXI (1926), 310.

Jesse, O. Die Höhe der leuchtenden Nachtwolken. *Astr. Nachr.,* CXL (1896), 161.

————Untersuchungen über die sogenannten leuchtenden Nachtwolken. *Sitzungsberichte der Kg. Preus. Akademie der Wissenshaften,* II (1890), 103.

Junge, C. *Met. Zeit.,* LIII (1936), 186-8.

————*Gerlands Beitr. z. Geophys.,* XLVI (1935), 108-29.

Kaemtz, L. F. *Vorlesungen über Meteorologie.* Halle (1841), 144-52.

Kelvin (Lord). *Proc. Roy. Soc. Edin.* (Feb., 1870).

Köhler, H. Uber das Irisieren und einige andere Erscheinungen in den Wolken. *Met. Zeit.,* XLIV (1929).

————Zur Condensation des Wasserdampfes in der Atmosphäre. *Geofys. Publ.,* II (1921, No. 1; 1922, No. 6).

————On Water in the Clouds. *Ibid.,* V (1927, No. 1).

————The Nucleus and the Growth of Hygroscopic Droplets. *Transactions of the Faraday Society,* XXXII (Aug., 1936, Part 8).

Kopp, W. Auslösung von Fallstreifen im Acu-Niveau durch einfallende Ci-Fallstreifenmittels Aer Obs. Lindenberg, 1928.

Kotsch, W. J. An Example of Colloidal Instability of Clouds in Tropical Latitudes. *Bull. Am. Met. Soc.,* XXVIII (Feb., 1947) 87-9.

Küttner, J. Moazagotl und Föhnwelle. *Beitr, z. Phys. d. f. Atm.,* XXV (1938), 79; XXV (1939), 251.

Landsberg, H. Atmospheric Condensation Nuclei. *Gerlands Beiträge, Ergebnisse der kosmischen Physik,* III, 155-252.

Langille, R. C. *et al.* S-Band Radar Echoes from Snow. *Report 26, Canadian Army Operational Research Group* (June 14, 1945).

————and K. L. S. Gunn. Quantitative Analysis of Vertical Structure in Precipitation. *J. Met.,* V (Dec., 1948), 301-4.

Langmuir, Irving, Vincent Schaefer, *et al. Final Report Project Cirrus.* Report No. RL 140, General Electric Research Laboratory, Schenectady, 1948.

————*Meteorological Research. Ibid.,* 1947.

Laufer, Maurice K. and Laurence W. Foskett. The Daytime Photo-electric Measurement of Cloud Heights. *Journal of the Aeronautical Sciences,* VIII (March, 1941).

Letzmann, J. Kinematik des Lenticularis und Castellatus. *Met. Zeit.,* L (1933), 365.

Löhner, H. Condensation Trails. *Luftwissen,* VII (1940), 337.

Ludlam, F. H. The Forms of Ice Clouds. *Q. J. Roy. Met. Soc.,* LXXIV (Jan., 1948), 39-56.

Malsch. Wolkenbildung durch Meteore. *Zeitschrift für angenehme Meteorologie,* L (1933), 325.

Marshall, J. S., L. G. Eon, and L. G. Tibbles. Analysis of Storm Echoes in Height using MHF. *Report 30, Canadian Ormy Operational Research Group.*

————R. C. Langille, and W. M. Palmer. Measurement of Rain by Radar. *J. Met.,* IV (Dec., 1947), 186-92.

————Summer Storm Echoes on Radar MEW, *Report 18, Canadian Army Operational Research Group* (Nov. 27, 1944).

Melander, G. *Sur la Condensation de la Vapeur d'eau dans l'atmosphère.* Helsingfors, 1897.

Mézin, M. Remarques sur le Film des Evolutions du Ciel Réalisé par H. R. Condit. *La Météorologie* (juillet- septembre, 1946), 337-40.

Middleton, W. E. K. *Meteorological Instruments.* 2nd ed., Toronto: University of Toronto Press, 1943.

————On the Theory of the Ceiling Projector. *Journal of the Optical Society of America,* XXIX (Aug., 1933), 340–9.

Miller, Robert W. The Use of Airborne Navigational and Bombing Radars for Weather Radar Operations and Verifications. *Bull. Am. Met. Soc.* (Jan., 1947), 19-28.

Möhn, H. *Met. Zeit.* (March, 1893), 81-97.

Nolan, P. J. Experiments on Condensation Nuclei. *Proc. Roy. Irish Acad.,* XLVII (Oct., 1941), Section A, No. 2.

Orr, J. L., D. Fraser, and K. G. Pettit. Analysis of Experiments on Inducing Precipitation. *National Research Council Report No. MD-32* (Aug. 17, 1949).

Owens, J. S. Sea Salt and Condensation Nuclei. *Q. J. Roy. Met. Soc.,* LXVI (Jan., 1940).

Palmer, H. P. Natural Ice-Particle Nuclei. *Ibid.,* LXXV (Jan., 1949), 15-22.

Perrie, D. W. The Rain Required for a Radar Echo. *Bull. Am. Met. Soc.,* XXX (Oct., 1949), 278-81.

————Cloud Seeding Experiments, November 1948. *Technical Circular Number 54,* Meteorological Division, Department of Transport, Canada (Jan., 1949).

————Induced Precipitation. *The New Trail,* VII (May, 1949), 100-5. Alumni Association, University of Alberta.

Petterssen, S. Condensation Caused by Mixing. *Q. J. Roy. Met. Soc.,* LXVIII (April, 1942), 167-73.

————*Weather Analysis and Forecasting.* New York: McGraw-Hill 1940.

Renou, E. Instructions Météorologiques. *Annuaire de la Société Météorologique de France,* III, Paris, 1855.

Rink, J. Moazagotls Wetterwolke. *Met. Zeit.,* LIV (1937), 190.

Robinson, G. D. The Distribution of Electricity in Thunderclouds. *Q. J. Roy Met. Soc.,* LXVII (Oct., 1941), 332–40.

Ryde, J. W. Echo intensities and attenuation due to clouds, rain, hail, sand and dust storms at centimetre wavelengths. Research Laboratories of the

General Electric Company (British) (Oct. 13, 1941).

Schmauss, A. and A. Wigand. *Die Atmosphäre als Kolloid.* Braunschweig, 1929.

Schmidt, W. Zur Erklärung der gesetzmässigen Verteilung der Tropfengrössen bei Regenfällen. *Met. Zeit.,* XXV (1908).

Sekara, Z. Helmholtz waves in a linear temperature field with vertical wind shear. *J. Met.,* V (June, 1948), 93-102.

Simon, A. Contribution a l'étude des Nuages. *Bulletin de l'Institute d'Egypte,* XXVI (1943–1944).

Simpson, G. C. On the Formation of Cloud and Rain. *Q. J. Roy. Met. Soc.,* LXVII (April, 1941), 99–133.

————Coronae and Iridescent Clouds. *Ibid.,* XXXVIII (Oct., 1912), 291.

————Sea-salt and Condensation Nuclei. *Ibid.,* LXVII (April, 1941), 163–9.

————Sea-salt and Condensation Nuclei. *Ibid.,* LXV (Oct., 1939), 553.

————The Electricity of Cloud and Rain. *Ibid.,* LXVIII (Jan., 1942), 1–34.

————and G. D. Robinson. *Proc. Roy. Soc.,* London, (A) CLXXVII, 281–329.

————and F. J. Scrase. *Ibid.,* CLXI, 309–52.

Stickley, A. R. An evaluation of the Bergeron-Findeisen precipitation theory. *Monthly Weather Review,* LXVIII, 272–80.

Störmer, C. *Geofys. Publ.,* IV (1932, No. 4), 1–27.

————Height and Velocity of Luminous Night Clouds Observed in Norway, 1932. *Publication No. 6.* Oslo: University Observatory, 1933.

————*Geofys. Publ.,* V (1927, No. 2), 1–8.

Stüve, G. Zur Kenntnis Kristallisation des Wasserdampfes aus der Luft. *Gerl. Beitr.* XXXII (1931).

Süring, R. *Die Wolken.* Leipzig: Akademische Verlagsgesellschaft, Becker & Erler, 1941.

Thomson, A. Mother-of-Pearl Clouds. *J. Roy. Astron. Soc. Can.,* XXVI (Dec., 1932), 437–41.

United States Weather Bureau. *Manual of Cloud Forms and Codes for States of the Sky.* Circular "S", 2nd ed., Washington, D.C.: U. S. Govt. Printing Office.

Vassy, E. Commentaire sur la Cinématographie de l'Ensemble du Ciel. *La Météorologie* (juillet-septembre, 1946), 292–301.

Vegard, L. Investigations of the Auroral Spectrum. *Geophys. Publ.,* XL (1933, No. 4), 53.

Vestine, E. H. Noctilucent Clouds. *J. Roy. Astron. Soc. Can.* (July-Aug., Sept., 1934).

Walker, G. T. Clouds and Cells. *Procès Verbaux de l'Association de Météorologie, International Union of Geodesy and Geophysics* (Lisbon, 1933).

————Some Recent Work on Cloud Forms. *Q. J. Roy. Met. Soc.,* LXV (Jan., 1939), 28–30.

Watanabe, M., Y. Isimaru, and K. Yosinari. *Report on the Cloud Observations made at the Mara Meteorological Observatory.* Tokyo: Central Meteorological Office, 1931.

Wegener, A. *Thermodynamik der Atmosphäre.* Leipzig, 1911.

————Untersuchungen der Wolkenelemente auf dem Höhen Sonnblick (3106 m.). *Sitzber. d. math. naturw. Kl. d. Akad. d. Wiss.,* CXVII, Wien, 1908.

————*Met. Zeit.*, LXI (1926), 103.

Weickmann, Helmut. Elements of Cloud and Precipitation. Field Information Agency Final Report 1063, item No. 27 (Office of U. S. Military Government for Germany, presumably Berlin).

Wexler, R. Radar Detection of a Frontal Storm. *J. Met.*, IV (June, 1946), 38–44.

————and D. M. Swingle. Radar Storm Detection. *Bull. Am. Met. Soc.*, XXVIII (April, 1947), 163.

Whipple, F. J. W. How are Mock Suns Produced? *Q. J. Roy. Met. Soc.*, LXVI (July, 1940), 275–9.

————and J. A. Chalmers. On Wilson's Theory of the Collection of Charge by Falling Drops. *Ibid.*, LXX (April, 1944), 103–19.

Wigand, A. Die elektrokolloiden Eigenschaften der Atmosphäre. *Met. Zeit.*, XLVI (1929).

————*Met. Zeit.*, XXX (1913), 10–18.

————and E. Frankenberger. Die elektrostatische Stabilisierung von Nebel und Wolken und die Niederschlagsbildung. *Annalen der Hydr. u. Mar. Met.* X/XI, Berlin, 1931.

————Uber Beständigkeit und Koagulation von Nebel und Wolken. *Physikalische Zeitschrift,* XXXI (1930), 204–15.

Wilson, C. T. R. *Journal of the Franklin Institute,* CCVIII (1929), 1–12.

————Condensation of water vapour in the presence of dust-free air and other gases. *Phil. Trans. Roy. Soc.,* London, (A) CLXXXIX (1897), 265–307.

————On the condensation nuclei produced in gases by the action of Roentgen rays, uranium rays, ultraviolet light and other agents. *Ibid.,* (A) CXCII (1892), 403–53.

————On the comparative efficiency as condensation nuclei of positively and negatively charged ions. *Ibid.,* (A) CXCIII (1900), 289–308.

Wright, H. L. Atmospheric Opacity. *Q. J. Roy. Met. Soc.,* LXV (July, 1939), 411–42.

————Sea-salt Nuclei. *Ibid.,* LXVI (Jan., 1940), 3–11.

————The Origin of Sea-salt Nuclei. *Ibid.,* 11.

INDEXES

Subject Index

Index of Names